科技英语阅读与翻译

Reading and Translation of Science English

主　编　蒋　勇
副主编　冯雪红
编　委　（按姓氏笔画排列）
毛艳文　冯雪红　李　娟　蒋　勇

东南大学出版社
SOUTHEAST UNIVERSITY PRESS
·南京·

内 容 提 要

本书涵盖了数学、物理、化学、计算机与信息工程、能源科学、农学、植物学、机械工程、电气工程、石油化学工程、汽车工程、通信工程、环境工程等18个基础学科和工程领域，为学习者提供了较全面且选择性广的阅读素材。每篇文章都加以注释，课后附有练习，便于学习者检查学习效果并巩固所学内容。学习者通过学习能具备阅读科学与工程英语文献资料的能力。

本书可作为英语、商务英语等专业的本科、高职高专以及成人教育的教学用书，也可供非英语专业学生作为拓展课程教材使用。

图书在版编目（CIP）数据

科技英语阅读与翻译 / 蒋勇主编. — 南京 : 东南大学出版社, 2021.2(2024.12重印)

ISBN 978-7-5641-9453-6

Ⅰ.①科… Ⅱ.①蒋… Ⅲ.①科学技术-英语-阅读教学-教材 ②科学技术-英语-翻译-教材 Ⅳ.①N43

中国版本图书馆CIP数据核字(2021)第027677号

责任编辑：刘 坚　**责任校对**：张万莹　**封面设计**：王 玥　**责任印制**：周荣虎

科技英语阅读与翻译

Keji Yingyu Yuedu Yu Fanyi

编　　著	蒋 勇
出版发行	东南大学出版社
社　　址	南京市四牌楼2号(邮编:210096　电话:025-83793330)
经　　销	全国各地新华书店
印　　刷	广东虎彩云印刷有限公司
开　　本	787mm×1092mm　1/16
印　　张	12.75
字　　数	305千字
版　　次	2021年2月第1版
印　　次	2024年12月第3次印刷
书　　号	ISBN 978-7-5641-9453-6
定　　价	58.00元

前言 Preface

为了提高毕业生的就业竞争力，很多设置英语专业的高校在培养英语专业学生掌握扎实的语言技能的基础上，还通过一些专门用途英语课程的学习使学生初步掌握一定的专业知识，根据社会需求培养复合型英语人才。在课程设置上，科技英语方向是一种选择。复合型英语专业的应用型特征很明显，就是要通过科技英语课程的学习掌握有关科学技术的基本知识，提高英语专业学生的科技素养，使复合型英语人才受到用人单位的欢迎，在社会竞争中立于不败之地。为了达到这一目标，我们编写了此书，其主要特点如下：

1. 题材全面，内容新颖。本书所选材料覆盖了基础科学和应用科学的多个领域，既涵盖了经典科学如数学、物理，也包括了数字技术等领域的最新发展，融知识性与趣味性于一体。

2. 知识学习与语言训练相结合。每个单元有两篇文章，文章末有注释，并编写了多项练习，既有检查阅读理解的题型，也有词汇术语及段落翻译的实践性题目，通过完成练习，学习者可以在理解的基础上达到巩固知识和提高技能的目的。

3. 选择性、实用性强。科技英语课程通常设置在专业选修类，各个学校分配的课时不同，本书的题材广泛，内容充实，可以根据课时选择所学单元，并且练习的题型多样，可满足理解性与巩固性等不同学习目的。

本书主要用作英语、商务英语专业的本科、高职高专以及成人教育的教学用书，也可作为非英语专业学生作为拓展课程教材使用。

本书科技英语阅读部分第一到第五单元由蒋勇编写，第六到第十一单元由冯雪红编写，第十二到第十八单元由李娟编写，科技英语翻译部分由蒋勇编写，毛艳文负责校对。

本书的编写得到了常州工学院的支持，被列为校级教材建设项目，外国语学院院长李静教授提出了许多宝贵意见并给予大力支持，编者在此表示衷心的感谢。

由于时间紧，再加上编者水平有限，书中难免有疏漏之处，恳请读者批评指正。

本书附有课件，可通过扫描下面或封底的二维码下载，也可从东南大学出版社网站上"读者服务"栏目中下载。

编者

2021.1

目录 Contents

科技英语阅读

科技英语阅读

Unit 1 Mathematics

Text A Mathematics

Mathematics is the study of topics such as quantity (numbers), structure, space, and change. There is a range of views among mathematicians and philosophers as to the exact scope and definition of mathematics.

Mathematicians seek out patterns and use them to formulate new conjectures①. Mathematicians resolve the truth or falsity of conjectures by mathematical proof. When mathematical structures are good models of real phenomena, then mathematical reasoning can provide insight or predictions about nature. Through the use of abstraction② and logic, mathematics developed from counting, calculation, measurement, and the systematic study of the shapes and motions of physical objects. Practical mathematics has been a human activity for as far back as written records exist. The research required to solve mathematical problems can take years or even centuries of sustained inquiry.

Mathematics is essential in many fields, including natural science, engineering, medicine, finance and the social sciences. Applied mathematics has led to entirely new mathematical disciplines, such as statistics and game theory. Mathematicians also engage in pure mathematics, or mathematics for its own sake, without having any application in mind. There is no clear line separating pure and applied mathematics, and practical applications for what began as pure mathematics are often discovered.

When civilization began to trade, a need to count was created. When humans traded goods, they needed a way to count the goods and to calculate the cost of those goods. The very first device for counting numbers was the human hand, counting on fingers. To count beyond ten fingers, mankind used natural markers, rocks or shells. From that point, counting boards and the abacus were invented.

Carl Friedrich Gauss, known as the prince of mathematicians Gauss referred to mathematics as "the Queen of the Sciences". In the original Latin *Regina Scientiarum*, as well as in German *Königin der Wissenschaften*, the word corresponding to science means a "field of knowledge", and this was the original meaning of "*science*" in English, also; mathematics is in this sense a field of knowledge. The specialization restricting the meaning of "science" to natural science follows the rise of Baconian science, which contrasted "natural science" to scholasticism, the Aristotelean method of inquiring from first principles. The role of empirical experimentation③ and observation is negligible in mathematics, compared to natural sciences such as biology, chemistry, or physics. Albert Einstein stated that "as far as the laws of mathematics refer to reality, they are not certain; and as far as they are certain, they do not refer to reality." More recently, Marcus du Sautoy has called mathematics "the Queen of Science ... the main driving force behind scientific discovery".

Many philosophers believe that mathematics is not experimentally falsifiable, and thus not a science according to the definition of Karl Popper④. However, in the 1930s Gödel's incompleteness theorems⑤ convinced many mathematicians that mathematics cannot be reduced to logic alone.

An alternative view is that certain scientific fields (such as theoretical physics) are mathematics with axioms⑥ that are intended to correspond to reality. The theoretical physicist J. M. Ziman proposed that science is public knowledge, and thus includes mathematics. Mathematics shares much in common with many fields in the physical sciences, notably the exploration of the logical consequences of assumptions. Intuition and experimentation also play a role in the formulation of conjectures in both mathematics and the (other) sciences. Experimental mathematics continues to grow in importance within mathematics, and computation and simulation are playing an increasing role in both the sciences and mathematics.

The opinions of mathematicians on this matter are varied. Many mathematicians feel that to call their area a science is to downplay the importance of its aesthetic side, and its history in the traditional seven liberal arts; others feel that to ignore its connection to the sciences is to turn a blind eye to the fact that the interface between mathematics and its applications in science and engineering has driven much development in mathematics. One way this difference of viewpoint plays out is in the philosophical debate as to whether mathematics is created (as in art) or discovered (as in science).

In contemporary education, mathematics education is the practice of teaching and learning mathematics, along with the associated scholarly research.

Researchers in mathematics education are primarily concerned with the tools, methods and approaches that facilitate practice or the study of practice; however, mathematics education research, known on the continent of Europe as the didactics or pedagogy of mathematics, has developed into an extensive field of study, with its own concepts, theories, methods, national and international organisations, conferences and literature.

Elementary mathematics was part of the education system in most ancient civilisations, including Ancient Greece, the Roman Empire, Vedic society and ancient Egypt. In most cases, a formal education was only available to male children with a sufficiently high status, wealth or caste.

In Plato's division of the liberal arts into the trivium and the quadrivium, the quadrivium included the mathematical fields of arithmetic and geometry. This structure was continued in the structure of classical education that was developed in medieval Europe. Teaching of geometry was almost universally based on Euclid's Elements. Apprentices to trades such as masons, merchants and money-lenders could expect to learn such practical mathematics as was relevant to their profession.

The first mathematics textbooks to be written in English and French were published by Robert Recorde, beginning with *The Grounde of Artes* in 1540. However, there are many different writings on mathematics and mathematics methodology that date back to 1800 BCE. These were mostly located in Mesopotamia where the Sumerians were practicing multiplication and division. There are also artifacts demonstrating their own methodology for solving equations like the quadratic equation⑦. After the Sumerians some of the most famous ancient works on mathematics come from Egypt in the form of the Rhind Mathematical Papyrus and the Moscow Mathematical Papyrus. The more famous Rhind Papyrus has been dated to approximately 1650 BCE but it is thought to be a copy of an even older scroll. This papyrus was essentially an early textbook for Egyptian students.

In the Renaissance, the academic status of mathematics declined, because it was strongly associated with trade and commerce. Although it continued to be taught in European universities, it was seen as subservient to the study of Natural, Metaphysical and Moral Philosophy.

This trend was somewhat reversed in the seventeenth century, with the University of Aberdeen creating a Mathematics Chair in 1613, followed by the Chair in Geometry being set up in University of Oxford in 1619 and the Lucasian Chair of Mathematics⑧ being established by the University of Cambridge in 1662. However, it was uncommon for mathematics to be taught outside of the universities. Isaac Newton, for example, received no formal mathematics teaching until he joined Trinity College, Cambridge in 1661.

In the 18th and 19th centuries, the industrial revolution led to an enormous increase in urban populations. Basic numeracy skills, such as the ability to tell the time, count money and carry out simple arithmetic, became essential in this new urban lifestyle. Within the new public education systems, mathematics became a central part of the curriculum from an early age.

By the twentieth century, mathematics was part of the core curriculum in all developed countries.

Throughout most of history, standards for mathematics education were set locally, by individual schools or teachers, depending on the levels of achievement that were relevant to, realistic for, and considered socially appropriate for their pupils.

In modern times, there has been a move towards regional or national standards, usually under the umbrella of a wider standard school curriculum. In England, for example, standards for mathematics education are set as part of the National Curriculum for England, while Scotland maintains its own educational system. In the USA, the National Governors Association⑨ Center for Best Practices and the Council of Chief State School Officers have published the national mathematics Common Core State Standards Initiative.

Ma (2000)⑩ summarised the research of others who found, based on nationwide data, that students with higher scores on standardised mathematics tests had taken more mathematics courses in high school. This led some states to require three years of mathematics instead of two. But because this requirement was often met by taking another lower level mathematics course, the additional courses had a "diluted" effect in raising achievement levels.

In North America, the National Council of Teachers of Mathematics has published the Principles and Standards for School Mathematics, which boosted the trend towards reform mathematics. In 2006, they released Curriculum Focal Points, which recommend the most important mathematical topics for each grade level through grade 8. However, these standards are enforced as American states and Canadian provinces choose.

Arguably the most prestigious award in mathematics is the Fields Medal⑪, established in 1936 and awarded every four years (except around World War II) to as many as four individuals. The Fields Medal is often considered a mathematical equivalent to the Nobel Prize.

The Wolf Prize in Mathematics⑫, instituted in 1978, recognizes lifetime achievement, and another major international award, the Abel Prize, was introduced in 2003. The Chern Medal was introduced in 2010 to recognize lifetime achievement. These accolades are awarded in recognition of a particular body of work, which may be innovational, or provide a solution to an outstanding problem in an established field.

A famous list of 23 open problems, called "Hilbert's problems"⑬, was compiled in 1900 by German mathematician David Hilbert. This list achieved great celebrity among mathematicians, and at least nine of the problems have now been solved. A new list of seven important problems, titled the "Millennium Prize Problems"⑭, was published in 2000. A solution to each of these problems carries a $ 1 million reward, and only one (the Riemann hypothesis) is duplicated in Hilbert's problems.

Notes:

① conjecture 猜测;推测

② abstraction 抽象化

③ empirical experimentation 实证研究

④ Karl Popper 卡尔·波普尔(1902—1994),当代西方最有影响力的哲学家之一,研究的范围甚广,涉及科学方法论、科学哲学、社会哲学、逻辑学等

⑤ Gödel's incompleteness theorems 哥德尔不完全性定理

⑥ axiom 公理

⑦ quadratic equation 二次方程

⑧ Lucasian Chair of Mathematics 卢卡斯数学教授席位

⑨ the National Governors Association(美国)全国州长协会

⑩ Ma X, 2000. A longitudinal assessment of antecedent course work in mathematics and subsequent mathematical attainment[J]. Journal of Educational Research, 94 (1):16-29.

⑪ Fields Medal 菲尔兹奖

⑫ Wolf Prize in Mathematics 沃尔夫数学奖是沃尔夫奖的一个奖项,它和菲尔兹奖被共同誉为数学界的最高荣誉

⑬ Hilbert's problems 希尔伯特问题是德国著名数学家希尔伯特在 1900 年 8 月巴黎国际数学家代表大会上提出的最重要的数学问题,分属数学基础、数论、代数和几何以及数学分析四大块

⑭ Millennium Prize Problems 千禧年大奖难题,又称世界七大数学难题,是七个由美国克雷数学研究所(Clay Mathematics Institute,CMI) 于 2000 年 5 月 24 日公布的数学猜想

Exercises

Ⅰ. Read each of the following statements carefully and decide whether it is true or false according to the text.

1. Mathematics is the study of topics such as quantity (numbers), space, and time. ( )
2. Mathematical modeling is used to explain abstract theories in the science of mathematics. ( )
3. Mathematics and physics share much in common. ( )
4. In contemporary education, more attention is paid to the practice of teaching and learning than the associated scholarly research. ( )
5. In the Renaissance, the academic status of mathematics increased because of the rise of trade and commerce. ( )
6. The first mathematics textbooks were published in ancient Egypt. ( )

Ⅱ. Answer the following questions according to the text.

1. How long is the history of practical mathematics?
2. Which branch of mathematics continues to grow in importance? Why?
3. What were the reasons for mathematics to become a central part of the curriculum in the 18th and 19th centuries?
4. Which award in mathematics is the most prestigious one in the world?
5. How many of the Hilbert's problems have been solved?

Ⅲ. Translate the following terms into their Chinese or English equivalents.

1. mathematical proof
2. practical mathematics
3. computation and simulation
4. National Curriculum for England
5. standardised mathematics tests
6. 博弈论
7. 理论物理学
8. 初等数学;基础数学
9. 乘法和除法
10. 理论数学

Ⅳ. Translate the paragraph into Chinese.

Mathematics is essential in many fields, including natural science, engineering, medicine, finance and the social sciences. Applied mathematics has led to entirely new mathematical disciplines, such as statistics and game theory. Mathematicians also engage in pure mathematics, or mathematics for its own sake, without having any application in mind. There is no clear line separating pure and applied mathematics, and practical applications for what began as pure mathematics are often discovered.

Text B Applied Mathematics

Applied mathematics is a branch of mathematics that deals with mathematical methods that find use in science, engineering, business, computer science, and industry. Thus, applied mathematics is a combination of mathematical science and specialized knowledge. The term "applied mathematics" also describes the professional specialty in which mathematicians work on practical problems by formulating and studying mathematical models①. In the past, practical applications have motivated the development of mathematical theories, which then became the subject of study in pure mathematics where abstract concepts are studied for their own sake. The activity of applied mathematics is thus intimately connected with research in pure mathematics.

Historically, applied mathematics consisted principally of applied analysis, most notably differential equations②; approximation theory③ (broadly construed, to include representations, asymptotic methods, variational methods, and numerical analysis); and applied probability. These areas of mathematics related directly to the development of Newtonian physics, and in fact, the distinction between mathematicians and physicists was not sharply drawn before the mid-19th century. This history left a pedagogical legacy in the United States: until the early 20th century, subjects such as classical mechanics were often taught in applied mathematics departments at American universities rather than in physics departments, and fluid mechanics may still be taught in applied mathematics departments. Quantitative finance is now taught in mathematics departments across universities and mathematical finance is considered a full branch of applied mathematics. Engineering and computer science departments have traditionally made use of applied mathematics.

Today, the term "applied mathematics" is used in a broader sense. It includes the classical areas noted above as well as other areas that have become increasingly important in applications. Even fields such as number theory that are part of pure mathematics are now important in applications (such as cryptography), though they are not generally considered to be part of the field of applied mathematics *per se*④. Sometimes, the term "applicable mathematics" is used to distinguish between the traditional applied mathematics that developed alongside physics and the many areas of mathematics that are applicable to real-world problems today.

There is no consensus as to what the various branches of applied mathematics are. Such categorizations are made difficult by the way mathematics and science change over time, and also by the way universities organize departments, courses, and degrees.

Many mathematicians distinguish between "applied mathematics," which is concerned with mathematical methods, and the "applications of mathematics" within science and engineering. A biologist using a population model and applying known mathematics would not be doing applied mathematics, but rather using it; however, mathematical biologists have posed problems that have stimulated the growth of pure mathematics. Mathematicians such as Poincaré and Arnold deny the existence of "applied mathematics" and claim that there are only "applications of mathematics." Similarly, non-mathematicians blend applied mathematics and applications of mathematics. The use and development of mathematics to solve industrial problems is also called "industrial mathematics".

The success of modern numerical mathematical methods and software has led to the emergence of computational mathematics, computational science, and computational engineering, which use high-performance computing for the simulation of phenomena and the solution of problems in the sciences and engineering. These are often considered interdisciplinary.

Historically, mathematics was most important in the natural sciences and engineering. However, since World War II, fields outside of the physical sciences have spawned the creation of new areas of mathematics, such as game theory⑤ and social choice theory⑥, which grew out of economic considerations.

The advent of the computer has enabled new applications: studying and using the new computer technology itself (computer science) to study problems arising in other areas of science (computational science) as well as the mathematics of computation (for example, theoretical computer science, computer algebra, numerical analysis). Statistics is probably the most widespread mathematical science used in the social sciences, but other areas of mathematics, most notably economics, are proving increasingly useful in these disciplines.

Applied mathematics is closely related to other mathematical sciences.

Scientific Computing

Scientific computing includes applied mathematics (especially numerical analysis), computing science (especially high-performance computing), and mathematical modelling in a scientific discipline.

Computer Science

Computer science relies on logic, algebra, and combinatorics.

Operations Research and Management Science

Operations research and management science are often taught in faculties of engineering, business, and public policy.

Statistics

Applied mathematics has substantial overlap with the discipline of statistics. Statistical theorists study and improve statistical procedures with mathematics, and statistical research often raises mathematical questions. Statistical theory relies on probability and decision theory, and makes extensive use of scientific computing, analysis, and optimization; for the design of experiments, statisticians use algebra and combinatorial design. Applied mathematicians and statisticians often work in a department of mathematical sciences (particularly at colleges and small universities).

Actuarial Science

Actuarial science applies probability, statistics, and economic theory to assess risk in insurance, finance and other industries and professions.

Mathematical Economics

Mathematical economics is the application mathematical methods to represent theories and analyze problems in economics. The applied methods usually refer to nontrivial mathematical techniques or approaches. Mathematical economics is based on statistics, probability, mathematical programming (as well as other computational methods), operations research, game theory, and some methods from mathematical analysis. In this regard, it resembles (but is distinct from) financial mathematics, another part of applied mathematics.

According to the Mathematics Subject Classification (MSC), mathematical economics falls into the applied mathematics/other classification of category 91:

Game theory, economics, social and behavioral sciences with MSC2010 classifications for "game theory" at codes 91Axx and for "mathematical economics" at codes 91Bxx.

Other Disciplines

The line between applied mathematics and specific areas of application is often blurred. Many universities teach mathematical and statistical courses outside of the respective departments, in departments and areas including business, engineering, physics, chemistry, psychology, biology, computer science, scientific computation, and mathematical physics.

Notes:

① A mathematical model is a description of a system using mathematical concepts and language. Mathematical models are used in the natural sciences and engineering disciplines, as well as in the social sciences. 数学模型

② differential equation 微分方程

③ approximation theory 近似理论

④ *per se*: a Latin phrase meaning "by itself" or "in itself".

⑤ The study of mathematical models of conflict and cooperation between intelligent rational decision-makers. Game theory is mainly used in economics, political science, and psychology, as well as logic, computer science, biology and poker. 博弈论

⑥ A theoretical framework for analysis of combining individual opinions, preferences, interests, or welfares to reach a collective decision or social welfare in some sense. A non-theoretical example of a collective decision is enacting a law or set of laws under a constitution. 社会选择理论

Exercises

Work in groups and discuss the following questions.

1. What is the main concern of applied mathematics?
2. What is the relationship between applied mathematics and Newtonian physics?
3. Why does applied mathematics have broader sense nowadays?
4. What new applications have been brought out with the invention of computers?
5. Explain the application of mathematics in mathematical economics.

Unit 2 Physics

Text A The Science of Matter, Space and Time

Have you ever wondered how often you could split a grain of sand into smaller pieces? Have you asked yourself what the sky is made of? Perhaps you have dreamed of travelling backwards in time?

Physicists are as curious as you are. They look for answers to questions that people have pondered since they first began to wonder about the world and their place within it. It often seems that for every answer physicists find, two new questions arise.

Exploring the Nature of Nature

Particle physicists try to understand the nature of nature at the smallest scales possible. Today, we know that atoms do not represent the smallest unit of matter. Particles called quarks and leptons seem to be the fundamental building blocks[①]—but perhaps there is something even smaller. Physicists are still far from understanding why a proton has about 2,000 times more mass than an electron. And on top of it all scientists suspect a whole new class of undiscovered supersymmetric particles[②] to complete the subatomic family.

Empty space, we have discovered, is actually not empty at all. Quantum effects constantly produce particles and antiparticles "out of nothing," only to have them disappear few moments later. And space itself can either be almost flat or curved, depending on the amount of matter it contains.

We have also learned that many subatomic processes can be reversed in time—but not every process. There are some small but crucial differences in the way matter and antimatter behave. Could it be the reason why our universe is made of matter, while antimatter has all but disappeared?

Astrophysicists have found that less than 10 percent of the mass of the entire universe consists of the kind of "luminous" matter that we can see. What is the dark matter that makes up the rest of the universe? How can we find out? Though we understand many important properties of the fundamental building blocks of our universe, there are untold mysteries still to solve.

Advances in technology allow physicists to build more powerful and sophisticated instruments to look deeper and deeper inside matter. Like adventurers entering unknown territory, physicists forge ahead into ever smaller dimensions.

What will be their next discovery?

- What is the world made of?
- How to find the smallest particles
- What to expect in the future
- Why support science
- Worldwide discoveries that established the Standard Model

What Is the World Made of?

The building blocks

Physicists have identified 12 building blocks that are the fundamental constituents of matter. Our everyday world is made of just three of these building blocks: the up quark, the down quark and the electron. This set of particles is all that's needed to make protons and neutrons and to form atoms and molecules. The electron neutrino observed in the decay of other particles, completes the first set of four building blocks.

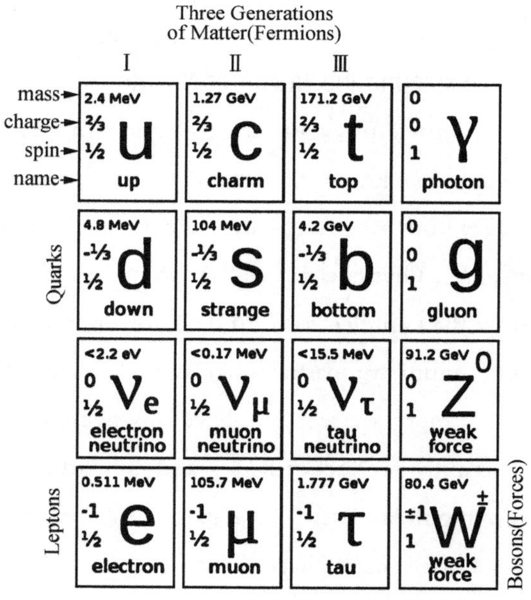

The building blocks of matter

For some reason nature has elected to replicate this first generation of quarks and leptons to produce a total of six quarks and six leptons, with increasing mass. Like all quarks, the sixth quark, named top, is much smaller than a proton (in fact, no one knows how small quarks are), but the top is as heavy as a gold atom!

Although there are reasons to believe that there are no more sets of quarks and leptons, theorists speculate that there may be other types of building blocks, which may partly account for the dark matter implied by astrophysical observations. This poorly understood matter exerts gravitational forces and manipulates galaxies. It will take earth-based accelerator experiments to identify its fabric.

The Building Blocks of Nature

The forces

Scientists distinguish four elementary types of forces acting among particles: strong, weak, electromagnetic and gravitational force.

- The strong force is responsible for quarks to be sticking together to form protons, neutrons and related particles.
- The electromagnetic force binds electrons to atomic nuclei (clusters of protons and neutrons) to form atoms.
- The weak force facilitates the decay of heavy particles into smaller siblings.
- The gravitational force acts between massive objects. Although it plays no role at the microscopic level, it is the dominant force in our everyday life and throughout the universe.

Particles transmit forces among each other by exchanging force-carrying particles called bosons③. These force mediators④ carry discrete amounts of energy, called quanta, from one particle to another. You could think of the energy transfer due to boson exchange as something like the passing of a basketball between two players.

Each force has its own characteristic bosons:

- The gluon mediates the strong force; it glues quarks together.
- The photon carries the electromagnetic force; it also transmits light.
- The W and Z bosons represent the weak force; they introduce different types of decays.

Physicists expect that the gravitational force may also be associated with a boson particle named the graviton. This hypothetical boson is extremely hard to observe since, at the subatomic level, the gravitational force is many orders of magnitude weaker than the other three elementary forces.

Notes:

① building block 原义指积木，此处指物质的组成部分

② supersymmetric particle 超对称粒子

③ 玻色子是遵循玻色-爱因斯坦统计，自旋量子数为整数的粒子，源自印度物理学家萨特延德拉·纳特·玻色（Satyendra Nath Bose）。

④ force mediator 作用力介质

Exercises

Ⅰ. Read each of the following statements carefully and decide whether it is true or false according to the text.

1. Atoms are the smallest units that make up for all matter. (　　)
2. Particle scientists have found all particles in the subatomic family. (　　)
3. Astrophysicists are very certain that the dark matter exists in the dark part of the universe. (　　)
4. The Standard Model has described all the elementary particles that make up all matter. (　　)
5. The strong force is responsible for the actions between massive objects. (　　)
6. The gluon carries the electromagnetic force; it also transmits light. (　　)

Ⅱ. Answer the following questions according to the text.

1. What types of particles are the fundamental building blocks of nature?
2. Is the seemingly empty space actually empty? Why/Why not?
3. How many types of forces are acting among particles? What are they?
4. How do particles transmit forces among each other?
5. Why is the hypothetical boson extremely hard to observe?

Ⅲ. Translate the following terms into their Chinese or English equivalents.

1. particle physics
2. subatomic family
3. particles and antiparticles
4. quantum effects
5. proton/neutron/molecule
6. 物质与反物质
7. 天体物理学
8. 地心引力

9. 原子核

10. 引力子

Ⅳ. Translate the paragraph into Chinese.

Particles transmit forces among each other by exchanging force-carrying particles called bosons. These force mediators carry discrete amounts of energy, called quanta, from one particle to another. You could think of the energy transfer due to boson exchange as something like the passing of a basketball between two players. Physicists expect that the gravitational force may also be associated with a boson particle named the graviton. This hypothetical boson is extremely hard to observe since, at the subatomic level, the gravitational force is many orders of magnitude weaker than the other three elementary forces.

Text B End of the Smashed Phone Screen? Self-healing Glass Discovered by Accident

By Samuel Gibbs

New type of polymer glass that can mend itself when pressed together is in development by University of Tokyo after a student discovered it.

Japanese researchers say they have developed a new type of glass that can heal itself from cracks and breaks.

Glass made from a low weight polymer called "polyether-thioureas" can heal breaks when pressed together by hand without the need for high heat to melt the material.

The research, published in *Science*, by researchers led by Professor Takuzo Aida from the University of Tokyo, promises healable glass that could potentially be used in phone screens and other fragile devices, which they say are an important challenge for sustainable societies.

While self-healing rubber and plastics have already been developed, the researchers said that the new material was the first hard substance of its kind that can be healed at room temperature.

"High mechanical robustness① and healing ability tend to be mutually exclusive," wrote the researchers, saying that while some hard but healable materials have been developed, "in most cases, heating to high temperatures, on the order of② 120 ℃ or more, to reorganise their cross-linked networks, is necessary for the fractured③ portions to repair."

The new polymer glass is "highly robust mechanically yet can readily be repaired by compression at fractured surfaces".

The properties of the polyether-thioureas glass④ were discovered by accident by graduate school student Yu Yanagisawa, who was preparing the material as a glue. Yanagisawa found that when the surface of the polymer was cut the edges would adhere to each other, healing to form a strong sheet after being manually compressed for 30 seconds at 21 ℃.

Further experimentation found that the healed material regained its original strength after a couple of hours.

Yanagisawa told NHK that he didn't believe the results at first and repeated his experiments multiple times to confirm the finding. He said: "I hope the repairable glass becomes a new environment-friendly material that avoids the need to be thrown away if broken."

This is not the first time a polymer has been suggested as a healable screen for devices such as smartphones. Researchers at the University of California proposed the use of polymer that could stretch to 50 times its original size and heal breaks within 24 hours.

Smartphone manufacturers have already used self-healing materials in devices. LG's G Flex 2 shipped in 2015 with a coating on its back that was capable of healing minor scratches over time, although failed to completely repair heavier damage.

According to research commissioned by repair firm iMend⑤ in 2015, over 21% of UK smartphone users were living with a broken screen, with smashed displays being one of the biggest issues alongside poor battery life.

A Japanese researcher has developed—by accident—a new type of glass that can be repaired simply by pressing it back together after it cracks.

The discovery opens the way for super-durable glass that could triple the lifespan of everyday products like car windows, construction materials, fish tanks and even toilet seats.

Yu Yanagisawa, a chemistry researcher at the University of Tokyo, made the breakthrough by chance while investigating adhesives that can be used on wet surfaces.

Does this mean you will soon be able to repair those cracks in your smartphone with a quick press of the fingers? Or surreptitiously piece together a shattered beer glass dropped after one pint too many?

Well, not quite. Not now and in fact, not in the near future.

But it does open a window of opportunity for researchers to explore ways to make more durable, lightweight, glass-like items, like car windows.

In a lab demonstration for AFP, Yanagisawa broke a glass sample into two pieces.

He then held the cross sections of the two pieces together for about 30 seconds until the glass repaired itself, almost resembling its original form.

To demonstrate its strength, he then hung a nearly full bottle of water from the piece of glass—and it stayed intact.

The organic glass, made of a substance called polyether thioureas, is closer to acrylic than mineral glass, which is used for tableware and smartphone screens.

Other scientists have demonstrated similar properties by using rubber or gel materials but Yanagisawa was the first to demonstrate the self-healing concept with glass.

The secret lies in the thiourea, which uses hydrogen bonding to make the edges of the shattered glass self-adhesive, according to Yanagisawa's study.

But what use is all this if it cannot produce a self-healing smartphone screen?

"It is not realistically about fixing what is broken, more about making longer-lasting resin glass," Yanagisawa told AFP.

Glass products can fracture after years of use due to physical stress and fatigue.

"When a material breaks, it has already had many tiny scars that have accumulated to result in major destruction," Yanagisawa said.

"What this study showed was a path toward making a safe and long-lasting resin glass", which is used in a wide range of everyday items.

"We may be able to double or triple the lifespan of something that currently lasts for 10 or 20 years," he said.

Notes:

① robustness 稳健性,鲁棒性

② on the order of: of around or about (a specified number); approximately 大约,差不多

③ fracture: if something hard fractures or is fractured, it breaks or cracks 破裂,断裂

④ polyether-thioureas glass 可自修复愈合的硬质聚合物玻璃;自愈玻璃

⑤ iMend 英国一家根据电话上门维修移动电话和平板电脑的维修公司

Exercises

Work in groups and discuss the following questions.

1. What substance makes it possible for glass having cracks and breaks to heal itself?
2. What is the difference between self-healing rubber, self-healing plastics and healable glass?
3. Who discovered the repairable properties of the polyether-thioureas glass?
4. Was the invention of the polyether-thioureas glass the only and the first case to use a polymer in a healable screen?
5. What was the result of the research commissioned by repair firm iMend in 2015?

Unit 3 Chemistry

Text A What Is Chemistry?

By Mary Bagley

Chemistry is the study of matter, its properties, how and why substances combine or separate to form other substances, and how substances interact with energy. Many people think of chemists as being white-coated scientists mixing strange liquids in a laboratory, but the truth is we are all chemists.

Doctors, nurses and veterinarians must study chemistry, but understanding basic chemistry concepts is important for almost every profession. Chemistry is part of everything in our lives.

Every material in existence is made up of matter—even our own bodies. Chemistry is involved in everything we do, from growing and cooking food to cleaning our homes and bodies to launching a space shuttle. Chemistry is one of the physical sciences that help us to describe and explain our world.

Five Branches

There are five main branches of chemistry, each of which has many areas of study.

Analytical chemistry① uses qualitative and quantitative observation to identify and measure the physical and chemical properties of substances. In a sense, all chemistry is analytical.

Physical chemistry combines chemistry with physics. Physical chemists study how matter and energy interact. Thermodynamics② and quantum mechanics③ are two of the important branches of physical chemistry.

Organic chemistry specifically studies compounds that contain the element carbon. Carbon has many unique properties that allow it to form complex chemical bonds and very large molecules. Organic chemistry is known as the "Chemistry of Life" because all of the molecules that make up living tissue have carbon as part of their makeup.

Inorganic chemistry studies materials such as metals and gases that do not have carbon as part of their makeup.

Biochemistry is the study of chemical processes that occur within living organisms.

Fields of Study

Within these broad categories are countless fields of study, many of which have important effects on our daily life. Chemists improve many products, from the food we eat and the clothing we wear to the materials with which we build our homes. Chemistry helps to protect our environment and searches for new sources of energy.

Food Chemistry

Food science deals with the three biological components of food — carbohydrates, lipids and proteins. Carbohydrates are sugars and starches, the chemical fuels needed for our cells to function. Lipids are fats and oils and are essential parts of cell membranes④ and to lubricate and cushion organs within the body. Because fats have 2.25 times the energy per gram than either carbohydrates or proteins, many people try to limit their intake to avoid becoming overweight. Proteins are complex molecules composed of from 100 to 500 or more amino acids⑤ that are chained together and folded into three-dimensional shapes necessary for the structure and function of every cell. Our bodies can synthesize some of the amino acids; however eight of them, the essential amino acids, must be taken in as part of our food. Food scientists are also concerned with the inorganic components of food such as its water content, minerals, vitamins and enzymes.

Food chemists improve the quality, safety, storage and taste of our food. Food chemists may work for private industry to develop new products or improve processing. They may also work for government agencies such as the Food and Drug Administration to inspect food products and handlers to protect us from contamination or harmful practices. Food chemists test products to supply information used for the nutrition labels or to determine how packaging and storage affects the safety and quality of the food. Flavorists⑥ work with chemicals to change the taste of food. Chemists may also work on other ways to improve sensory appeal, such as enhancing color, odor or texture.

Environmental Chemistry

Environmental chemists study how chemicals interact with the natural environment. Environmental chemistry is an interdisciplinary[⑦] study that involves both analytical chemistry and an understanding of environmental science. Environmental chemists must first understand the chemicals and chemical reactions present in natural processes in the soil water and air. Sampling and analysis can then determine if human activities have contaminated the environment or caused harmful reactions to affect it.

Water quality is an important area of environmental chemistry. "Pure" water does not exist in nature; it always has some minerals or other substance dissolved in it. Water quality chemists test rivers, lakes and ocean water for characteristics such as dissolved oxygen, salinity, turbidity[⑧], suspended sediments, and pH. Water destined for human consumption must be free of harmful contaminants and may be treated with additives like fluoride and chlorine to increase its safety.

Agricultural Chemistry

Agricultural chemistry is concerned with the substances and chemical reactions that are involved with the production, protection and use of crops and livestock. It is a highly interdisciplinary field that relies on ties to many other sciences. Agricultural chemists may work with the Department of Agriculture, the Environmental Protection Agency, the Food and Drug Administration or for private industry. Agricultural chemists develop fertilizers, insecticides and herbicides necessary for large-scale crop production. They must also monitor how these products are used and their impacts on the environment. Nutritional supplements are developed to increase the productivity of meat and dairy herds.

Agricultural biotechnology is a fast-growing focus for many agricultural chemists. Genetically manipulating crops to be resistant to the herbicides used to control weeds in the fields requires detailed understanding of both the plants and the chemicals at the molecular level. Biochemists must understand genetics, chemistry and business needs to develop crops that are easier to transport or that have a longer shelf life.

Chemical Engineering

Chemical engineers research and develop new materials or processes that involve chemical reactions. Chemical engineering combines a background in chemistry with engineering and economics concepts to solve technological problems. Chemical engineering jobs fall into two main groups: industrial applications and development of new products.

Industries require chemical engineers to devise new ways to make the manufacturing of their products easier and more cost effective. Chemical engineers are involved in designing and operating processing plants, develop safety procedures for handling dangerous materials, and supervise the manufacture of nearly every product we use. Chemical engineers work to develop new products and processes in every field from pharmaceuticals to fuels and computer components.

Geochemistry

Geochemists combine chemistry and geology to study the makeup and interaction between substances found in the Earth. Geochemists may spend more time in field studies than other types of chemists. Many work for the US Geological Survey or the Environmental Protection Agency in determining how mining operations and waste can affect water quality and the environment. They may travel to remote abandoned mines to collect samples and perform rough field evaluations, and then follow a stream through its watershed to evaluate how contaminants are moving through the system. Petroleum geochemists are employed by oil and gas companies to help find new energy reserves. They may also work on pipelines and oil rigs to prevent chemical reactions that could cause explosions or spills.

Notes:

① analytical chemistry 分析化学

② thermodynamics 热力学,热动力学

③ quantum mechanics 量子力学

④ cell membrane 细胞膜

⑤ amino acid 氨基酸

⑥ flavorist: also known as flavor chemist, is someone who uses chemistry to engineer artificial and natural flavors 调香师

⑦ interdisciplinary 跨学科的, 跨领域的

⑧ turbidity 浊度,混浊度

Exercises

Ⅰ. Read each of the following statements carefully and decide whether it is true or false according to the text.

1. Chemistry is a science studied only by the chemists and it is beyond the reach of ordinary

people. ( )

2. The five main branches of chemistry are analytical chemistry, physical chemistry, organic chemistry, inorganic chemistry and materials chemistry. ( )
3. Carbohydrates, lipids and proteins are the chemical fuels needed for our cells to function. ( )
4. Environmental chemists study how chemicals interact with the natural environment. ( )
5. Chemical engineers research and develop new materials or processes that involve physical reactions. ( )
6. The main location for Geochemists is field. ( )

Ⅱ. Answer the following questions according to the text.

1. Why is the understanding of basic chemistry concepts important for all?
2. What are the main branches of chemistry?
3. Explain the function of carbohydrates.
4. Why are many people quite concerned about their intake of lipids?
5. What are the water characteristics that are used to judge water quality?

Ⅲ. Translate the following terms into their Chinese or English equivalents.

1. physical science
2. qualitative and quantitative observation
3. suspended sediments
4. nutritional supplements
5. lipids
6. 碳水化合物
7. 水污染;水体污染
8. 化学反应
9. 化学键
10. 生物体;生命体

Ⅳ. Translate the paragraph into Chinese.

Geochemists combine chemistry and geology to study the makeup and interaction between substances found in the Earth. Geochemists may spend more time in field studies than other types of chemists. Many work for the US Geological Survey or the Environmental Protection Agency in determining how mining operations and waste can affect water quality and the environment. They may travel to remote abandoned mines to collect samples and perform rough field evaluations, and then follow a stream through its watershed to evaluate how contaminants are moving through the system.

Text B New Clues Could Help Scientists Harness the Power of Photosynthesis

Summary: A discovery has been made that could enable scientists to design better ways to use light energy and to engineer crop plants that more efficiently harness the energy of the sun. The identification of a gene needed to expand light harvesting in photosynthesis into the far-red-light spectrum provides clues to the evolution of oxygen-producing photosynthesis, an evolutionary advance that changed the history of life on Earth.

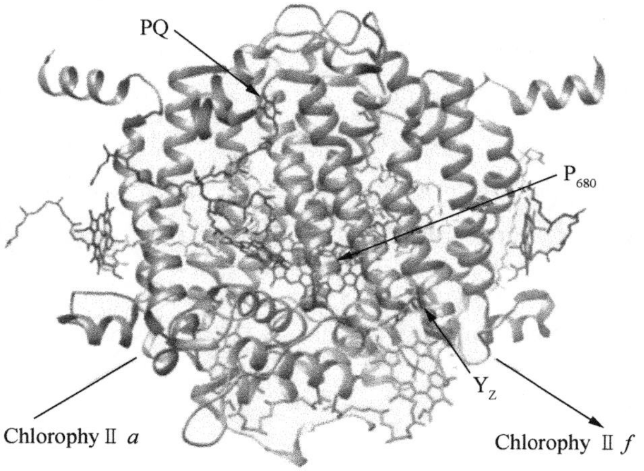

This illustration is a model of Chl f synthase, potentially a ChlF dimer, based on the known X-ray structure of the core of the Photosystem Ⅱ reaction center. Photosystem Ⅱ is the light-driven enzyme that oxidizes water to produce oxygen in plants, algae and cyanobacteria. The bright green molecules near the bottom of the structure are imagined to be substrate Chl a molecules that would be oxidized in light to produce Chl f.

Credit: Donald A. Bryant lab, Penn State University

Identification of a gene needed to expand light harvesting in photosynthesis into the far-red-light spectrum provides clues to the development of oxygen-producing photosynthesis, an evolutionary advance that changed the history of life on Earth. "Knowledge of how photosynthesis evolved could empower scientists to design better ways to use light energy for the benefit of humankind," said Donald A. Bryant, the Ernest C. Pollard Professor of

Biotechnology and professor of biochemistry and molecular biology at Penn State University and the leader of the research team that made the discovery.

This discovery, which could enable scientists to engineer① crop plants that more efficiently harness the energy of the Sun, will be published online by the journal *Science*.

"Photosynthesis usually ranks about third after the origin of life and the invention of DNA in lists of the greatest inventions of evolution," said Bryant. "Photosynthesis was such a powerful invention that it changed Earth's atmosphere by producing oxygen, allowing diverse and complex life forms—algae, plants, and animals—to evolve."

The researchers identified the gene that converts chlorophyll a—the most abundant light-absorbing pigment used by plants and other organisms that harness energy through photosynthesis—into chlorophyⅡ *f*—a type of chlorophyⅡ that absorbs light in the far-red range of the light spectrum. There are several different types of chlorophyⅡ, each tuned to absorb light in different wavelengths. Most organisms that get their energy from photosynthesis use light in the visible range, wavelengths of about 400 to 700 nanometers. Bryant's lab previously had shown that chlorophyⅡ *f* allows certain cyanobacteria—bacteria that use photosynthesis and that are sometimes called blue-green algae—to grow efficiently in light just outside of the usual human visual range—far-red light (700 to 800 nanometers). The ability to use light wavelengths other than those absorbed by plants, algae, and other cyanobacteria② confers a powerful advantage to those organisms that produce chlorophyⅡ *f*—they can survive and grow when the visible light they normally use is blocked.

"There is nearly as much energy in the far-red and near-infrared light that reaches the Earth from the Sun as there is in visible light," said Bryant. "Therefore, the ability to extend light harvesting in plants into this range would allow the plants to more efficiently use the energy from the Sun and could increase plant productivity."

The gene the researchers identified encodes an enzyme that is distantly related to one of the main components of the protein machinery used in oxygen-producing photosynthesis. The researchers showed that the conversion of chlorophyⅡ a to chlorophyⅡ *f* requires only this one enzyme in a simple system that could represent an early intermediate stage in the evolution of photosynthesis. Understanding the mechanism by which the enzyme functions could provide clues that enable scientists to design better ways to use light energy.

"There is intense interest in creating artificial photosynthesis as an alternative energy source," said Bryant. "Understanding the evolutionary trajectory③ that nature used to create oxygen production in photosynthesis is one component that will help scientists design an efficient and effective system. The difficulty is that photosynthesis is an incredibly complex process with hundreds of components and, until now, there were few known intermediate stages in its evolution. The simple system that we describe in this paper provides a model that can be further manipulated experimentally for studying those early stages in the evolution of photosynthesis."

By disabling the gene that encodes the enzyme in two cyanobacteria that normally produce chlorophyⅡ *f*, the researchers demonstrated that the enzyme is required for the production of chlorophyⅡ *f*. The experiment showed that, without this enzyme, these cyanobacteria could no longer synthesize chlorophyⅡ *f*. By artificially adding the gene that encodes the enzyme, the researchers also showed that this one enzyme is all that is necessary to convert cyanobacteria that normally do not produce chlorophyⅡ *f* into ones that can produce it.

Another clue that the newly identified enzyme could represent an early stage in the evolution of photosynthesis is that the enzyme requires light to catalyze its reaction and may not require oxygen, as scientists had previously suspected. "Because the enzyme that synthesizes chlorophyⅡ *f* requires light but may not require oxygen for its activity, it is possible that it evolved before Photosystem Ⅱ, the photosynthetic complex that produces oxygen and to which the enzyme is related. If the enzyme is an evolutionary predecessor of Photosystem Ⅱ, then evolution borrowed an enzyme that was originally used for chlorophyⅡ synthesis and used it to evolve an enzyme that could produce oxygen, which ultimately led to changes in Earth's atmosphere," said Bryant.

Notes:

① engineer: design and build 设计

② cyanobacterium: (pl. cyanobacteria) 藻青菌

③ evolutionary trajectory 进化轨迹

Exercises

Work in groups and discuss the following questions.

1. Give the list of the greatest inventions of evolution.
2. What is the function of chlorophyⅡ *f*?
3. Have scientists learned the process of photosynthesis? Why/Why not?
4. How does the newly identified enzyme work in the evolution of photosynthesis?
5. Why is the ability to use light wavelengths other than those absorbed by plants, algae, and other cyanobacteria significant?

Unit 4 Mechanical Engineering

Text A Mechanical Engineering Overview

Mechanical engineering is one of the largest, broadest, and oldest engineering disciplines. Mechanical engineers use the principles of energy, materials, and mechanics to design and manufacture machines and devices of all types. They create the processes and systems that drive technology and industry.

The key characteristics of the profession are its breadth, flexibility, and individuality①. The career paths of mechanical engineers are largely determined by individual choices, a decided advantage in a changing world.

Mechanics, energy and heat, mathematics, engineering sciences, design and manufacturing form the foundation of mechanical engineering. Mechanics includes fluids, ranging from still water to hypersonic gases flowing around a space vehicle; it involves the motion of anything from a particle to a machine or complex structure.

Mechanical engineers research, design, develop, manufacture, and test tools, engines, machines, and other mechanical devices. Mechanical engineering is one of the broadest engineering disciplines. Engineers in this discipline work on power-producing machines such as electric generators, internal combustion engines, and steam and gas turbines. They also work on power-using machines such as refrigeration and air-conditioning equipment, machine tools②, material-handling systems, elevators and escalators, industrial production equipment, and robots used in manufacturing. Some mechanical engineers design tools that other engineers need for their work. In addition, mechanical engineers work in manufacturing or agriculture production, maintenance, or technical sales; many become administrators or managers.

The Field

Mechanics, energy and heat, mathematics, engineering sciences, design and manufacturing form the foundation of mechanical engineering. Mechanics includes fluids, ranging from still water to hypersonic gases flowing around a space vehicle; it involves the motion of anything from a particle to a machine or complex structure.

Analysis, design, and synthesis are the key functions of mechanical engineers. The question is often how devices and processes actually work. The first step is to visualize what is happening and clearly state the problem. A mechanical engineer will then use computer-based modeling, simulation, and visualization techniques to test different solutions.

Design is one of the most satisfying jobs for a mechanical engineer. To design and build a new car, you must reckon with③ power, weight, size and shape, materials, reliability, and safety. "Synthesis"④ is when you pull all the factors together in a design that can be successfully manufactured. Design problems are challenging because most are open-ended, without a single or best answer. There is no best mousetrap⑤— just better ones.

The field is notable for emphasizing versatility. A mechanical engineering education is an excellent foundation for work in other fields. Some mechanical engineers work on medical problems, such as the mechanics of bones and joints, or the fluid dynamics of the circulatory system. Mechanical engineers deal with economic issues, from the cost of a single component, to the economic impact of a manufacturing plant. M. E.'s can be found in sales, engineering management, and corporate management. Versatility is a decided asset in a world that is undergoing constant economic, political, industrial, and social change. Mechanical engineers are educated and positioned, not only to adapt, but to define and direct change.

The diversity of the field of mechanical engineering is represented in the following areas of involvement.

Basic Engineering

Fundamentally, mechanical engineers are involved with the mechanics of motion and the transfer of energy from one form to another or one place to another. Mechanical engineers design and build machines for industrial and consumer use—virtually any machine you find, had a mechanical engineer involved with its development and production. They design heating, ventilation, and air conditioning systems to control the climate in homes, offices, and industrial plants, and develop refrigeration systems for the food industry. Mechanical engineers also design heat exchangers, key components in high-tech mechanical and electronic computer equipment.

Energy Conversion

We live in a world of dependent on the production and conversion of energy into useful forms. Mechanical engineers are involved in all aspects of the production and conversion of energy from one form to another. We design and operate fossil fuel, hydroelectric, conventional, nuclear and cogeneration power plants[6]. We design and develop internal combustion engines for automobiles, trucks and marine use and also for electrical power generation.

Energy Resources

Mechanical engineers are experts on the conversion and use of existing energy sources and in developing the equipment needed to process and transport fuels. At the same time, mechanical engineers are active in finding and developing new forms of energy. In that effort, mechanical engineers deal with the production of energy from alternate sources, such as solar, geothermal, and wind.

Engineering & Technology Management

Working in project teams is a way of life for mechanical engineers. Deciding which projects to undertake and leading those projects to a successful conclusion is the job of experienced engineers who move into management. On the safety front, all projects involve safety issues. By its very nature mechanical engineering involves the harnessing and channeling[7] of the forces of nature, forces which are often extremely powerful. Consider the contained "explosion" that inflates an automobile air bag or the mechanical forces involved in bringing an airplane load of people to a safe and comfortable landing, or the safety and reliability of an elevator, a power plant, or an incubator for pre-maturely born infants.

Environment & Transportation

Transportation is a large and growing field for mechanical engineers. Existing modes of air and surface transport require continuous improvement or replacement. Mechanical engineers work at the cutting edge of these efforts. Wherever machines are made or used, you will find mechanical engineers. They are instrumental in the design, development and manufacturing of machines that transmit power. They are also critically involved with the environmental impact and fuel efficiency of the machines they develop and with any by-products of the fuels used to power those machines.

Manufacturing

In contemporary manufacturing companies, mechanical engineers play a key role in the "realization" of products, working closely with other engineers and specialists in corporate management, finance, marketing, and packaging. Mechanical engineers design products, select materials and processes, and convert them to finished products. They design and manufacture machine tools—literally the machines that make machines and design entire manufacturing processes, aided by the latest technologies in automation and robotics. Finally, the finished products are transported in equipment designed by mechanical engineers. This is the largest area of employment for mechanical engineers, especially when the process and textile industries are included. A finished product requires the right materials, a viable plant and equipment, and a manufacturing system. This all comes within the purview of[8] mechanical, manufacturing and industrial engineers.

Materials & Structures

In order to arrive at the best design for a product, mechanical engineers use a wide variety of metal, plastic, ceramic materials. They also use composites made up of more than one type of material. Once designed, built and in service, elements like pipeline welds and sections, gears and other drive-train elements may need inspection for structural integrity or the effects of mechanical wear. Non-Destructive Evaluation, as its name implies, allows mechanical engineers to use X-ray, magnetic particle, ultrasound and other techniques to examine the internal condition of structural and machine parts, without causing them to fail or without removing them from service. This analysis is particularly important in assuring the reliability and safety of pressure vessels and piping systems.

Systems & Design

Most mechanical engineers work in the design and control of mechanical, electromechanical and fluid power systems. As a mechanical engineer functioning as a design engineer it is likely that you would be involved with one or more technical specialties, for example: Robotic System Design; Computer Coordinated Mechanisms; Expert Systems in Design; Computer-Aided Engineering; Geometric Design; Design Optimization; Kinematics and Dynamics of Mechanisms; Cam Design/Gear Design; Power Transmission; or Design of Machine Elements. Design engineers take into account a truly wide number of factors in the course of their work, such as: product performance, cost, safety, manufacturability, serviceability, human factors, aesthetic appearance, durability, reliability, environmental impact and recyclability.

Notes:

① breadth, flexibility, and individuality 广度、灵活性与个性化

② machine tool 机床

③ reckon with: take into account 将……加以考虑

④ synthesis 综合,协同

⑤ mousetrap 引人注目的新产品

⑥ cogeneration power plant 联合发电厂

⑦ channel 输送,传送

⑧ within the purview of 在……范围之内

Exercises

Ⅰ. Read each of the following statements carefully and decide whether it is true or false according to the text.

1. Mechanical engineering is one of the largest, broadest, and oldest engineering disciplines. ( )
2. Some mechanical engineers design tools that other engineers need for their work. ( )
3. Fluids, one branch of mechanics, range from flowing water to hypersonic gases flowing around a space vehicle. ( )
4. Analysis, manufacture, and synthesis are the key functions of mechanical engineers. ( )
5. Mechanical engineering is notable for emphasizing specialization. ( )
6. Mechanical engineering and economic issues are separated and have nothing to do with each other. ( )

Ⅱ. Answer the following questions according to the text.

1. What will determine the career paths of mechanical engineers?
2. Give the list of the foundation of mechanical engineering.
3. What do mechanical engineers usually do in their filed of discipline?

4. What is the first step in order to make devices and processes actually work?

5. Why are synthesis and design problems challenging?

Ⅲ. Translate the following terms into their Chinese or English equivalents.

1. engineering disciplines
2. machine tools
3. visualization techniques
4. fossil fuel
5. alternate sources
6. 发电机
7. 企业管理;公司管理
8. 内燃机
9. 燃气轮机
10. 通排风系统;换气系统

Ⅳ. Translate the paragraph into Chinese.

Fundamentally, mechanical engineers are involved with the mechanics of motion and the transfer of energy from one form to another or one place to another. Mechanical engineers design and build machines for industrial and consumer use—virtually any machine you find, had a mechanical engineer involved with its development and production. They design heating, ventilation, and air conditioning systems to control the climate in homes, offices, and industrial plants, and develop refrigeration systems for the food industry. Mechanical engineers also design heat exchangers, key components in high-tech mechanical and electronic computer equipment.

Text B Mechanical Engineering

Mechanical engineering is the discipline that applies the principles of engineering, physics, and materials science for the design, analysis, manufacturing, and maintenance of mechanical systems. It is the branch of engineering that involves the design, production, and operation of machinery. It is one of the oldest and broadest engineering disciplines.

The engineering field requires an understanding of core concepts including mechanics, kinematics, thermodynamics, materials science, structural analysis, and electricity. Mechanical engineers use these core principles along with tools like computer-aided design, and product lifecycle management to design and analyze manufacturing plants, industrial equipment and machinery, heating and cooling systems, transport systems, aircraft, watercraft, robotics, medical devices, weapons, and others.

Mechanical engineering emerged as a field during the Industrial Revolution in Europe in the 18th century; however, its development can be traced back several thousand years around the world. Mechanical engineering science emerged in the 19th century as a result of developments in the field of physics. The field has continually evolved to incorporate advancements in technology, and mechanical engineers today are pursuing developments in such fields as composites, mechatronics①, and nanotechnology. Mechanical engineering overlaps with aerospace engineering, metallurgical engineering, civil engineering, electrical engineering, manufacturing engineering, chemical engineering, industrial engineering, and other engineering disciplines to varying amounts. Mechanical engineers may also work in the field of biomedical engineering, specifically with biomechanics, transport phenomena, biomechatronics②, bionanotechnology, and modeling of biological systems.

Mechanical engineering finds its application in the archives of various ancient and medieval societies throughout mankind. In ancient Greece, the works of Archimedes (287 BC–212 BC) deeply influenced mechanics in the Western tradition and Heron of Alexandria (c. 10–70 AD)③ created the first steam engine (Aeolipile). In China, Zhang Heng (78–139 AD) improved a water clock and invented a seismometer, and Ma Jun (200–265 AD) invented a chariot with differential gears. The medieval Chinese horologist and engineer Su Song (1020–1101 AD) incorporated an escapement mechanism into his astronomical clock tower two centuries before any escapement can be found in clocks of medieval Europe, as well as the world's first known endless power-transmitting chain drive.

During the years from 7th to 15th century, the era called the Islamic Golden Age, there were remarkable contributions from Muslim inventors in the field of mechanical technology. Al-Jazari, who was one of them, wrote his famous *Book of Knowledge of Ingenious Mechanical Devices*④ in 1206, and presented many mechanical designs. He is also considered to be the inventor of such mechanical devices which now form the very basic of mechanisms, such as the crankshaft and camshaft.

Important breakthroughs in the foundations of mechanical engineering occurred in England during the 17th century when Sir Isaac Newton both formulated the three Newton's Laws of Motion and developed Calculus, the mathematical basis of physics. Newton was reluctant to publish his methods and laws for years, but he was finally persuaded to do so by his colleagues, such as Sir Edmund Halley⑤, much to the benefit of all mankind. Gottfried

Wilhelm Leibniz[6] is also credited with creating Calculus during the same time frame.

During the early 19th century in England, Germany and Scotland, the development of machine tools led mechanical engineering to develop as a separate field within engineering, providing manufacturing machines and the engines to power them. The first British professional society of mechanical engineers was formed in 1847 Institution of Mechanical Engineers, thirty years after the civil engineers formed the first such professional society Institution of Civil Engineers. On the European continent, Johann von Zimmermann (1820–1901) founded the first factory for grinding machines in Chemnitz, Germany in 1848.

In the United States, the American Society of Mechanical Engineers (ASME) was formed in 1880, becoming the third such professional engineering society, after the American Society of Civil Engineers (1852) and the American Institute of Mining Engineers (1871). The first schools in the United States to offer an engineering education were the United States Military Academy in 1817, an institution now known as Norwich University in 1819, and Rensselaer Polytechnic Institute in 1825. Education in mechanical engineering has historically been based on a strong foundation in mathematics and science.

The field of mechanical engineering can be thought of as a collection of many mechanical engineering science disciplines. Several of these subdisciplines which are typically taught at the undergraduate level are listed below, with a brief explanation and the most common application of each. Some of these subdisciplines are unique to mechanical engineering, while others are a combination of mechanical engineering and one or more other disciplines. Most work that a mechanical engineer does uses skills and techniques from several of these subdisciplines, as well as specialized subdisciplines. Specialized subdisciplines, as used in this article, are more likely to be the subject of graduate studies or on-the-job training than undergraduate research. Several specialized subdisciplines are discussed in this section.

Mechanics is, in the most general sense, the study of forces and their effect upon matter. Typically, engineering mechanics is used to analyze and predict the acceleration and deformation (both elastic and plastic) of objects under known forces (also called loads) or stresses. Subdisciplines of mechanics include:

- Statics, the study of non-moving bodies under known loads, how forces affect static bodies
- Dynamics (or kinetics), the study of how forces affect moving bodies
- Mechanics of materials, the study of how different materials deform under various types of stress

- Fluid mechanics, the study of how fluids react to forces
- Kinematics, the study of the motion of bodies (objects) and systems (groups of objects), while ignoring the forces that cause the motion. Kinematics is often used in the design and analysis of mechanisms
- Continuum mechanics, a method of applying mechanics that assumes that objects are continuous (rather than discrete)

Mechanical engineers typically use mechanics in the design or analysis phases of engineering. If the engineering project were the design of a vehicle, statics might be employed to design the frame of the vehicle, in order to evaluate where the stresses will be most intense. Dynamics might be used when designing the car's engine, to evaluate the forces in the pistons and cams as the engine cycles. Mechanics of materials might be used to choose appropriate materials for the frame and engine. Fluid mechanics might be used to design a ventilation system for the vehicle, or to design the intake system for the engine.

Mechatronics is a combination of mechanics and electronics. It is an interdisciplinary branch of mechanical engineering, electrical engineering and software engineering that is concerned with integrating electrical and mechanical engineering to create hybrid systems. In this way, machines can be automated through the use of electric motors, servo-mechanisms⑦, and other electrical systems in conjunction with special software. A common example of a mechatronics system is a CD-ROM drive. Mechanical systems open and close the drive, spin the CD and move the laser, while an optical system reads the data on the CD and converts it to bits. Integrated software controls the process and communicates the contents of the CD to the computer.

Robotics is the application of mechatronics to create robots, which are often used in industry to perform tasks that are dangerous, unpleasant, or repetitive. These robots may be of any shape and size, but all are preprogrammed and interact physically with the world. To create a robot, an engineer typically employs kinematics (to determine the robot's range of motion) and mechanics (to determine the stresses within the robot).

Robots are used extensively in industrial engineering. They allow businesses to save money on labor, perform tasks that are either too dangerous or too precise for humans to perform them economically, and to ensure better quality. Many companies employ assembly lines of robots, especially in Automotive Industries and some factories are so robotized that they can run by themselves. Outside the factory, robots have been employed in bomb disposal, space exploration, and many other fields. Robots are also sold for various residential applications, from recreation to domestic applications.

Mechanical engineers are constantly pushing the boundaries of what is physically possible in order to produce safer, cheaper, and more efficient machines and mechanical systems. Some technologies at the cutting edge of mechanical engineering are listed below (see also exploratory engineering).

Notes:

① mechatronics 机械电子学

② biomechatronics 生物机械电子学

③ Heron of Alexandria 海伦,古希腊数学家、力学家、机械学家,约公元62 年活跃于亚历山大时期,在那里教过数学、物理学等课程

④ *Book of Knowledge of Ingenious Mechanical Devices* 加扎里的《精巧机械装置的知识》一书在中世纪阿拉伯机械史研究中具有重要意义,书中叙述了六大类共 50 种他所设计的自动机械装置的制作方法和工作流程

⑤ Sir Edmund Halley 埃德蒙·哈利(1656—1742),英国第一个成功预言并解释彗星运动的天文学家

⑥ Gottfried Wilhelm Leibniz 戈特弗里德·威廉·莱布尼茨,德国哲学家、数学家,历史上少见的通才,被誉为 17 世纪的亚里士多德。在数学上,他和牛顿先后独立发现了微积分,而且他所使用的微积分的数学符号被更广泛地使用

⑦ servo-mechanism 伺服机构;随动系统

Exercises

Work in groups and discuss the following questions.

1. What tools are used by mechanical engineers to design and analyze manufacturing plants, industrial equipment and machinery, heating and cooling systems, transport systems, etc.?
2. When and where did mechanical engineering emerge as a field of discipline?
3. Who made remarkable contributions in the field of mechanical technology from 7th to 15th century? Give specific examples to illustrate this.
4. Which important breakthroughs occurred in England during the 17th century?
5. When did the development of machine tools lead mechanical engineering to develop as a separate field?

Unit 5 Electrical and Electronic Engineering

Text A Electrical Engineering

Electrical engineering is a field of engineering that generally deals with the study and application of electricity, electronics, and electromagnetism. This field first became an identifiable occupation in the latter half of the 19th century after commercialization of the electric telegraph, the telephone, and electric power distribution and use. Subsequently, broadcasting and recording media made electronics part of daily life. The invention of the transistor, and later the integrated circuit, brought down the cost of electronics to the point they can be used in almost any household object.

Electrical engineering has now subdivided into a wide range of subfields including electronics, digital computers, power engineering, telecommunications, control systems, radio-frequency engineering, signal processing, instrumentation, and microelectronics. The subject of electronic engineering is often treated as its own subfield but it intersects with all the other subfields, including the power electronics of power engineering.

Electrical engineers typically hold a degree in electrical engineering or electronic engineering. Practicing engineers may have professional certification and be members of a professional body. Such bodies include the Institute of Electrical and Electronics Engineers (IEEE) ①and the Institution of Engineering and Technology (IET) ②.

Electrical engineers work in a very wide range of industries and the skills required are likewise variable. These range from basic circuit theory to the management skills required of a project manager. The tools and equipment that an individual engineer may need are similarly variable, ranging from a simple voltmeter to a top end analyzer to sophisticated design and manufacturing software.

Electricity has been a subject of scientific interest since at least the early 17th century. A prominent early electrical scientist was William Gilbert③ who was the first to draw a clear distinction between magnetism and static electricity and is credited with establishing the term electricity.

In the 19th century, research into the subject started to intensify. Notable developments in this century include the work of Georg Ohm, who in 1827 quantified the relationship between the electric current and potential difference in a conductor, of Michael Faraday, the discoverer of electromagnetic induction in 1831, and of James Clerk Maxwell④, who in 1873 published a unified theory of electricity and magnetism in his treatise *Electricity and Magnetism*.

During these decades use of electrical engineering increased dramatically. In 1882, Thomas Edison switched on the world's first large-scale electric power network that provided 110 volts—direct current (DC)—to 59 customers on Manhattan Island in New York City. In 1884, Sir Charles Parsons⑤ invented the steam turbine allowing for more efficient electric power generation.

During the development of radio, many scientists and inventors contributed to radio technology and electronics. The mathematical work of James Clerk Maxwell during the 1850s had shown the relationship of different forms of electromagnetic radiation including possibility of invisible airborne waves (later called "radio waves"). In his classic physics experiments of 1888, Heinrich Hertz proved Maxwell's theory by transmitting radio waves with a spark-gap transmitter, and detected them by using simple electrical devices.

The invention of the transistor in late 1947 by William B. Shockley, John Bardeen, and Walter Brattain of the Bell Telephone Laboratories opened the door for more compact devices and led to the development of the integrated circuit in 1958 by Jack Kilby and independently in 1959 by Robert Noyce. Starting in 1968, Ted Hoff and a team at the Intel Corporation invented the first commercial microprocessor, which foreshadowed the personal computer.

Subdisciplines

Electrical engineering has many subdisciplines, the most common of which are listed below. Although there are electrical engineers who focus exclusively on one of these subdisciplines, many deal with a combination of them. Sometimes certain fields, such as electronic engineering and computer engineering, are considered separate disciplines in their own right.

Power

Power engineering deals with the generation, transmission, and distribution of electricity as well as the design of a range of related devices. These include transformers, electric generators, electric motors, high voltage engineering, and power electronics. In many regions of the world, governments maintain an electrical network called a power grid that connects a variety of generators together with users of their energy. Users purchase electrical energy from the grid, avoiding the costly exercise of having to generate their own. Power engineers may work on the design and maintenance of the power grid as well as the power systems that connect to it. Such systems are called on-grid power systems and may supply the grid with additional power, draw power from the grid, or do both. Power engineers may also work on systems that do not connect to the grid, called off-grid power systems, which in some cases are preferable to on-grid systems. The future includes Satellite controlled power systems, with feedback in real time to prevent power surges and prevent blackouts.

Control

Control systems play a critical role in space flight.

Control engineering focuses on the modeling of a diverse range of dynamic systems and the design of controllers that will cause these systems to behave in the desired manner. To implement such controllers, electrical engineers may use electronic circuits, digital signal processors, microcontrollers, and programmable logic controls (PLCs) [⑥]. Control engineering has a wide range of applications from the flight and propulsion systems of commercial airliners to the cruise control present in many modern automobiles. It also plays an important role in industrial automation.

Control engineers often utilize feedback when designing control systems. For example, in an automobile with cruise control the vehicle's speed is continuously monitored and fed back to the system which adjusts the motor's power output accordingly. Where there is regular feedback, control theory can be used to determine how the system responds to such feedback.

Electronics

Electronic engineering involves the design and testing of electronic circuits that use the properties of components such as resistors, capacitors, inductors, diodes, and transistors to achieve a particular functionality. The tuned circuit[⑦], which allows the user of a radio to filter out all but a single station, is just one example of such a circuit. Another example (of a pneumatic signal conditioner) is shown in the adjacent photograph.

Prior to the Second World War, the subject was commonly known as radio engineering and basically was restricted to aspects of communications and radar, commercial radio, and early television. Later, in post war years, as consumer devices began to be developed, the field grew to include modern television, audio systems, computers, and microprocessors. In the mid-to-late 1950s, the term radio engineering gradually gave way to the name electronic engineering.

Before the invention of the integrated circuit in 1959, electronic circuits were constructed from discrete components that could be manipulated by humans. These discrete circuits consumed much space and power and were limited in speed, although they are still common in some applications. By contrast, integrated circuits packed a large number—often millions—of tiny electrical components, mainly transistors, into a small chip around the size of a coin. This allowed for the powerful computers and other electronic devices we see today.

Microelectronics

Microelectronics engineering deals with the design and microfabrication of very small electronic circuit components for use in an integrated circuit or sometimes for use on their own as a general electronic component. The most common microelectronic components are semiconductor transistors, although all main electronic components (resistors, capacitors, etc.) can be created at a microscopic level. Nanoelectronics is the further scaling of devices down to nanometer levels. Modern devices are already in the nanometer regime, with below 100 nm processing having been standard since about 2002.

Microelectronic components are created by chemically fabricating wafers⑧ of semiconductors such as silicon (at higher frequencies, compound semiconductors like gallium arsenide and indium phosphide) to obtain the desired transport of electronic charge and control of current. The field of microelectronics involves a significant amount of chemistry and material science and requires the electronic engineer working in the field to have a very good working knowledge of the effects of quantum mechanics.

Signal processing deals with the analysis and manipulation of signals. Signals can be either analog, in which case the signal varies continuously according to the information, or digital, in which case the signal varies according to a series of discrete values representing the information. For analog signals, signal processing may involve the amplification and filtering of audio signals for audio equipment or the modulation and demodulation of signals for telecommunications.

For digital signals, signal processing may involve the compression, error detection and error correction of digitally sampled signals.

Signal Processing is a very mathematically oriented and intensive area forming the core of digital signal processing and it is rapidly expanding with new applications in every field of electrical engineering such as communications, control, radar, audio engineering, broadcast engineering, power electronics, and biomedical engineering as many already existing analog systems are replaced with their digital counterparts. Analog signal processing is still important in the design of many control systems.

DSP processor ICs are found in every type of modern electronic systems and products including, SDTV | HDTV⑨ sets, radios and mobile communication devices, Hi-Fi audio equipment, Dolby noise reduction algorithms, GSM mobile phones, mp3 multimedia players, camcorders and digital cameras, automobile control systems, noise cancelling headphones, digital spectrum analyzers, intelligent missile guidance, radar, GPS based cruise control systems, and all kinds of image processing, video processing, audio processing, and speech processing systems.

Notes:

① Institute of Electrical and Electronics Engineers 电气和电子工程师协会(IEEE)是一个国际性的电子技术与信息科学工程师的协会,是目前全球最大的非营利性专业技术学会,其会员人数超过 40 万人,遍布 160 多个国家和地区。IEEE 致力于电气、电子、计算机工程和与科学有关的领域的开发和研究,在太空、计算机、电信、生物医学、电力及消费性电子产品等领域已制定了 900 多个行业标准,现已发展成为具有较大影响力的国际学术组织

② Institution of Engineering and Technology 英国工程技术学会,简称 IET,成立于 1871 年,是工程技术领域的全球顶级专业学术学会,专业分类包括能源电力、交通运输、信息与通信、设计与制造、建筑环境 5 大行业 40 多个专业领域

③ William Gilbert 威廉·吉尔伯特(1544—1603),英国皇家科学院物理学家,主要在电学和磁力学方面有很大贡献

④ James Clerk Maxwell 詹姆斯·克拉克·麦克斯韦(1831—1879),出生于苏格兰爱丁堡,英国物理学家、数学家,经典电动力学的创始人,统计物理学的奠基人之一

⑤ Charles Parsons 查尔斯·帕森斯,19 世纪著名的科学家和发明家,他发明的帕森斯涡轮机有效地提高了轮船的速度

⑥ programmable logic controls 可编程逻辑控制器是一种专门为在工业环境下应用而设计的数字运算操作电子系统。它采用一种可编程的存储器,在其内部存储执行逻辑

运算、顺序控制、定时、计数和算术运算等操作的指令，通过数字式或模拟式的输入输出来控制各种类型的机械设备或生产过程

⑦ tuned circuit 对于包含电容和电感及电阻元件的无源一端口网络，其端口可能呈现容性、感性及电阻性。当电路端口的电压 U 和电流 I 出现同相位，电路呈电阻性时，称为谐振现象，这样的电路，称为谐振电路

⑧ wafer 晶圆，是指硅半导体集成电路制作所用的硅晶片，由于其形状为圆形，故称为晶圆；在硅晶片上可加工制作成各种电路元件结构，而成为有特定电性功能的集成电路产品

⑨ SDTV 标准清晰度电视，HDTV 高清晰度电视

Exercises

Ⅰ. Read each of the following statements carefully and decide whether it is true or false according to the text.

1. Electrical engineering first became an identifiable occupation in the latter half of the 18th century. (　　)
2. Electronic engineering is not a subfield of electrical engineering. (　　)
3. Voltmeters and analyzers are the tools engineers may need, while manufacturing software is not. (　　)
4. The research into electricity started from the early 17th century and started to intensify in the 19th century. (　　)
5. In 1884, Thomas Edison invented the steam turbine allowing for more efficient electric power generation. (　　)
6. James Clerk Maxwell had shown the relationship of different forms of electromagnetic radiation and had detected them by using simple electrical devices. (　　)

Ⅱ. Answer the following questions according to the text.

1. What reduced the cost of electronics to the point they can be used in almost any household object?
2. What is the relationship between electrical engineering and electronic engineering?
3. Explain the development of transistor, more compact devices and integrated circuits.
4. Why are control systems necessary?
5. What are the main objectives of microelectronics engineering?

Ⅲ. Translate the following terms into their Chinese or English equivalents.

1. electrical engineering
2. electronic engineering

3. control systems
4. Institution of Engineering and Technology
5. potential difference
6. 电报
7. 集成电路
8. 直流电流
9. 电磁辐射
10. 微处理器

Ⅳ. Translate the paragraph into Chinese.

Microelectronics engineering deals with the design and microfabrication of very small electronic circuit components for use in an integrated circuit or sometimes for use on their own as a general electronic component. The most common microelectronic components are semiconductor transistors, although all main electronic components (resistors, capacitors, etc.) can be created at a microscopic level. Nanoelectronics is the further scaling of devices down to nanometer levels. Modern devices are already in the nanometer regime, with below 100 nm processing having been standard since about 2002.

Text B Basic Electronics Concepts and Theory

One of the key basic concepts used in electronics is electrical current. Current is the amount of electrical charge that flows through a given area (like a wire) per unit time (second). Current is expressed in Amperes, which is generally referred to as Amps in practice or just "A" when writing a specific number of amps. When drawing the flow of current in a circuit they symbol "I" is used (uppercase i) with subscripts to denote which segment the current is flowing through. The technical definition of one amp is one coulomb of charge per second. Since charge is carried by electronics, it is very useful to think of current as the flow of electrons through a component or wire.

The electrical flow of current is a popular way of thinking and visualizing the flow of electrons, but the actual flow of electrons is reversed of the way we typically think (and draw) current moving through a circuit. When we say that current is flowing from point A to point B, we intuitively think that the electrons are flowing from point A with a higher voltage to point B with a lower voltage.

In fact the flow of electrons is the reverse. Electrons flow from the lower voltage at point B to the higher voltage at point A. The reason for this confusion is that when electricity was first being explored by scientists it was assumed that positive charges were moving around the circuit when in reality electrons with their negative charge move freely through conductors. The conventions① established based on a positive charge carrier was actually established by Benjamin Franklin well before the true nature of charge carriers was understood and is something we still use today. This wrong guess hundreds of years ago has caused great confusion for everyone that learns electronics ever since.

Luckily the historical blunder has very little impact on us today. While knowing the physics of electronics is important, especially to understand the physics of how semiconductor devices work, only a handful of areas and applications require that the physics be used directly. For the most part it is just easier to stick with the more intuitive assumption that accompanies the description of current flowing, like water, from point A to point B.

The visualization of current flowing like water is actually one of the best analogies for describing how electricity flows and is used to describe how many components actually work. In the water flowing analogy, current is actual water flow and the pressure on the water is equivalent to voltage in electronics. Resistance in water flow, such as a section of smaller diameter pipe in a run of pipe is also called resistance in electronics and has the same impact on current flow that a small diameter pipe has on the flow of water.

The flow of electric currents creates a number of effects, many of which are quite useful. Any current flowing in a wire will create a magnetic field. This effect has a large number of uses. The magnetic field created by current can be used to generate radio waves for standard AM/FM radio, cell phone signals, Bluetooth signals and every other wireless communication method. The magnetic field produced by a current flow is also used to power motors, actuators②, and otherwise move or pick up objects by carefully applying the produced magnetic field. The flow of current also heats up the conductor it flows through. This allows things like resistive heaters③ on an electric stove to be made and also is why your cell phone gets hot when you are on a call for a long time. This makes current and the magnetic field it produces one of the primary means of electronics interacting with the physical world.

Resistor

Resistors are one of the most common passive electrical components and are found in almost every electronic circuit. Resistor applications vary widely, from current limiting, heating, sensing, voltage drops, and feedback loops are a few common examples. A resistor limits the amount of current that can flow through a circuit based the resistance of the resistor. Electrical resistance is measured in ohms and can be determined through the use of Ohms Law which is one of the most fundamental equations in electrical engineering.

Electrical Resistance

The heart of a resistor is its electrical resistance. Electrical resistance can be visualized through an analogy of water moving through pipes. When water flowing through a pipe encounters a narrower section of pipe, the flow of the water slows due to the restriction of the smaller diameter pipe. In the analogy, the flow of water is equivalent to an electrical current and the narrower section of pipe is equivalent to a resistor.

When an electrical potential④ or voltage (equivalent to water pressure) encounters a resistor, the current running through the circuit is restricted. Electrical resistance is expressed as ohms (Ω) and depends on the resistance of the material, geometric shape, and length of the resistive path.

Resistors do have other characteristics of importance beyond their electrical resistance, particularly their power rating, voltage rating, temperature coefficient of resistance, thermal noise, capacitance⑤ and inductance⑥ of the resistor. These resistor properties can be very important to take in to account, depending upon the application.

Ohm's Law

One of the most fundamental equations used in electrical engineering is Ohm's Law. The relationship between voltage, current, and resistance was discovered by Georg Ohm in 1827. Ohm discovered that the drop in electrical potential (voltage) across a resistor was directly proportional to the current through the resistor or $I = V/R$ where I = current (amps), V = voltage (volts), and R = resistance (ohms). Ohm's Law can be solved for any of the variables yielding, $V = I \times R$, or the voltage across a resistor is equal to the current running through the resistor times the resistance of the resistor, and $R = V/I$, or resistance is equal to the voltage across the resistance divided by the current through the resistance. These equivalent formulas for Ohm's Law are some of the most used and most important equations used in circuit analysis, design, and electrical engineering.

Notes:

① convention 惯例;常规

② actuator (电磁铁)螺线管

③ resistive heater 电阻发热丝;电阻式热源

④ electrical potential 电位;电势

⑤ capacitance 电容;电容量;电容值

⑥ inductance 电感;电感值

Exercises

Work in groups and discuss the following questions.

1. When we say that current is flowing from point A to point B, is it the same that the electrons are flowing from point A with a higher voltage to point B with a lower voltage?
2. How is a magnetic field created in a wire?
3. When is heating effect useful with the flow of current in a conductor?
4. How is electrical resistance measured?
5. Explain Ohm's Law and give the equation of Ohm's Law.

Unit 6 Agriculture

Text A Labor Mobility and the Rural Exodus

The aggregate statistics of regional migration patterns are somewhat elusive. West Africa's population has increased four or five times since the late nineteenth century and urban centers now contain a fifth of this much-expanded total. This means that rural populations are now at least three times what they were. Some villages may have stayed roughly the same in numbers, with people trickling out gradually as a kind of homeostatic regulator①; Even so the majority of villagers have had to go and live elsewhere, mostly still in the countryside. Moreover, a skew in the age and sex distribution of those who move (among whom younger men predominate) can leave the local division of labor badly disorganized despite stability or even growth in the size of a given rural population. Suffice it to say that all this mobility in a period of extraordinary growth is both a significant economic resource (freeing labor as a commodity in large quantities) and a cause for some political disquiet.

There have been two great movements of population in West Africa. The earlier, which is over a hundred years old, is from the savannah interior to export-cropping zones nearer the coast, notably the forest belt. This rural-rural migration pattern is still important, but extremely difficult to quantify, because rates of natural population increase also vary widely within the region. The second movement, which is substantially a feature of post-World War II history, is the rapid growth of West African cities, especially capital cities. Annual growth rates for the region's urban centers have been consistently higher than overall growth rates of 3 percent, being 4–10 percent in various countries and 5–6 percent on the average for West Africa as a whole during the period since 1950. The number of West Africans now living in urban areas (30 million) is about six times what it was three decades ago, and the urban share of the population has more than tripled to its present level of 20 percent.

The concentration of this growth in a few centers of government is revealed by an index of primate urbanization based on the ratio of the first- and second-largest cities in each country. Thus in six countries (Senegal, Guinea, Guinea Bissau, Liberia, Mali, and Togo), the capital city was six to nine times bigger than the next-largest city at the beginning of the 1970s. In another four (the Gambia, Sierra Leone, the Ivory Coast, and Niger), it was four to five times bigger. And in five countries (Nigeria, Benin, Ghana, Upper Volta, and Mauritania), it was less than double the size of the second city. Of this last group, Mauritania② has an urban population that has been growing at the rate of 15 percent a year for over a decade; the capital city of Benin③ is not the largest city; Ghana is the most urbanized nation in West Africa; and Nigeria's several regions are all densely populated, with over half of West Africa's major cities distributed among them. In general, the tendency since decolonization has been for more migrants than ever before to make their destination a handful of national capitals.

Thus Dakar, which before the war had a population of 54,000, reached 800,000 by 1976; Lagos grew from 126,000 to 2.5 million in the same period; and Abidjan increased from 10,000 in 1931 to 555,000 in 1970. Less spectacularly, Monrovia went from 10,000 to 164,000 in forty years, and Accra's population increased a mere tenfold to 636,000 in 1970. Some of the smaller capital cities had the most dramatic increases: Niamey from 2,000 in 1931 to 108,000 in 1972 and Lome from 7,000 to 193,000 in the same period. Guinea's capital, Conakry, managed to grow to almost sixty times its original size between 1934 and 1972; from 9,000 to 526,000 (nine times the size of the next-largest city). These figures show that, in the post-World War II period, West Africa has acquired for the first time a genuine urban basis for its regional economy, one that has been fueled by massive growth in the total population and, in equal measure, by the migration of country dwellers from their villages to the new urban agglomerations that mark the rise of the postcolonial state.

The mobility of West Africa's population has thus been a highly visible feature of the region's modern history. Some would find in it a measure of instability and dislocation, mass exodus from the countryside being seen as an index of rural malaise. But this mobility is neither novel nor dysfunctional, for migration and movement were intrinsic to the indigenous population's way of life even before the concentration of economic opportunities on the coast set up today's asymmetrical drift from savannah villages to forest plantations and city slums. Warfare and slave raiding produced continuous population movements on a large and a small scale. Shifting agriculture allowed for micro-variations in settlement that over time could amount to substantial demographic change. Migration for political, religious, and trading purposes was common well before *the pax colonica*. Rules of exogamy created well-dispersed affined networks through which people could travel long distances. The development cycle of domestic groups routinely produced forces for residential fission that might occasion relocation either nearby or farther away. Migration was *not* encouraged among slaves and married women, who made their residential moves prior to assuming those statuses. For others, though it was sometimes fraught with danger, migration was normal, expected. I mention this as an antidote to any lingering belief that colonialism jerked a static indigenous population into a nomadic existence to which it was wholly unaccustomed.

There are two main types of long-distance migration. In one the migrant is a settler; this type is more likely if his destination is a rural setting, for he can readily establish there the conditions of reproducing himself and his family for the duration of his own lifetime and beyond. The other type is circulatory migration, in which the migrant expects to return to his home village sooner or later; this type may involve long or short stays in town or countryside. Labor circulation, a prominent feature of West African rural settings on a seasonal, annual, or longer-term basis, is the dominant mode of migration to the cities. The reason is simple: None but a few securely employed and affluent city dwellers can expect to find in urban areas the long-term life-support system they need. Residential flexibility and a way of life linking the migrant strongly to his home area are thus normal in West Africa. This means that the rural exodus, which has gathered momentum since 1960, should not necessarily be understood as the countryside's permanent loss of population. A shift in the balance of population in favor of the cities would occur if the rate of rural-urban circulation increased in the population as a whole and if the average length of absence from the village increased. It would not need to mean that the volume of new urban *settlers* had significantly increased, although it is undoubtedly true that the number of city dwellers without effective links to the countryside is growing all the time.

The mechanisms of mass migration both now and in the colonial period are a subject of considerable debate. Some writers, reacting against liberal models of rational decision making in a free labor market, have emphasized the elements of compulsion in modern West African migration—the use of force by colonial armies and recruiters, the imposition of money taxes, and pressure from village chiefs (today, famine, loss of land, and rural impoverishment might be enlisted under a similar rubric). Though there is a valid case to be made against neglect of these considerations, it is my opinion that the bulk of migration can be understood as having taken place within the framework of a 4 "free" labor market (in the legal rather than existential sense). Obviously the case of the Ivory Coast before World War I differs somewhat from that of southwestern Nigeria today in that regard; but if a general characterization has to be made, modern West African migratory movements have been largely voluntary.

This conclusion may affect our evaluation of the rural exodus. It leads me to describe the large picture as that of West Africans voting with their feet, if somewhat tentatively, for city over village life; others might understand the process as the ejection of an unwilling population from a rural setting that they would otherwise prefer.

One partial antidote to these polarized discussions about the causes of population movement is to stress the significance of studying rural social structures themselves to see how they differentially affect the release of labor. We have seen that traditional social life is not without its contradictions. Further studies may throw light on the reasons why some peoples lose most of their young adults while others retain them. Both the "liberal" and the "compulsion" theories of migration tend to treat rural West Africans as a passive, homogeneous mass, whereas rural society has its dynamics and variations that need to be taken fully into account.

It is also important to keep a historical time scale in mind when thinking about labor-migration trends. If the depression of 1930–45 is anything to go by, a prolonged slump in the world economy today will reverse the shift of people out of the countryside that occurred in the 1960s and early 1970s. Certainly, as jobs become harder to get, the possibility of migrating on a seasonal basis is much reduced, and this change may compel migrants to choose between staying longer abroad and never leaving the village at all. These long swings in the labor market affect the balance of power between laborers and employers. Thus, when a boom takes off, demand for labor exceeds supply, wage levels rise, work conditions are determined by what keeps the workers happy, and turnover and mobility are high. When there is a recession, labor supply exceeds demand, real wage levels fall, workers take any conditions they can get, and turnover and mobility are much reduced. We have a tendency sometimes to think of migration as a linear growth curve; the evidence of the last century in West Africa is that a long-term upward trend is masked by significant fluctuations in the medium term. The chances are that the early 1980s will constitute a downward fluctuation. To what extent is commercial agriculture responsible for the rural exodus[4]? Obviously, commercial agriculture, as in the classic examples of cocoa and groundnut farming, has played a significant part in drawing population away from the savannah hinterland to areas most suitable for export production. But this shift constitutes a redistribution of rural population, not a loss to the countryside as a whole.

There are two main ways in which it could be said that commercial agriculture has contributed to rural emigration. The first would be instances where concentration of capital, land, and technology has displaced rural workers and forced them to seek their livelihood elsewhere, let us say in the cities. As we saw in the previous chapter, there is scant evidence that any such development has yet taken place on a significant scale in West Africa. It is in another way that the commercial evolution of the countryside has encouraged emigration, and then only indirectly: Export-crop agriculture has been the economic basis for state aggrandizement⑤ in the region. The wholesale transfer of agricultural surpluses from rural areas to states whose expenditures are biased overwhelmingly in favor of urban areas is the main reason why more and more people make unfavorable economic comparisons between their own villages and the nation's principal cities. If income generated from agriculture stayed in the countryside there would be a much-reduced rural exodus. Country statistics show variation within West Africa from urban populations of less than 10 percent to several of more than 30 percent. Figures as large as the latter imply a massive alteration of national economic structure. Importation of cheap food from abroad has delayed the commercialization of food production for the home market that ought to accompany such a demographic shift. I shall argue later that urbanization on this scale presents enormous opportunities for the West African economies. But the prevailing opinion is that cities are dangerous and unhealthy places that represent both a drain on the public purse and a threat to the political status quo, to the extent that the swollen urban masses cannot find adequate work.

The fact is that the quality and quantity of material amenities are much higher in the cities than in most villages. Moreover, when people are concentrated in central places, markets are bigger and services can be delivered more cheaply on a per capita basis. Village life has its solid virtues, but economic horizons are necessarily limited there, and that limitation matters to a large number of people, especially young adults. The rulers of states, whose own economic behavior draws villagers in thousands to their capital cities, may sometimes regret their presence so close to the corridors of power, where they do more damage if they become upset. But they can hardly claim that these people would be better off to stay in the countryside, unless they match the claim with a massive redirection of state expenditures. Judgment on this issue is as much a matter of philosophy and politics as it is of the economic and demographic record.

Notes:

① homeostatic regulator 自我平衡的调节器

② Mauritania 毛里塔尼亚(西非国家,首都努瓦克肖特)

③ Benin 贝宁(位于非洲西部)

④ rural exodus 农村迁徙

⑤ state aggrandizement 国家强化

Exercises

Ⅰ. Read each of the following statements carefully and decide whether it is true or false according to the text.

1. The rural populations are now at least three times what they were in West Africa. (　　)
2. Shifting agriculture allowed for micro-variations in settlement that over time could not amount to substantial demographic change. (　　)
3. Commercial agriculture, as in the classic examples of cocoa and groundnut farming, has played a trivial part in drawing population away from the savannah hinterland to areas most suitable for export production. (　　)
4. The wholesale transfer of agricultural surpluses from rural areas to states whose expenditures are biased overwhelmingly in favor of urban areas is the main reason why more and more people make unfavorable economic comparisons between their own villages and the nation's principal cities. (　　)
5. Importation of cheap food from abroad has delayed the commercialization of food production for the home market that ought to accompany such a demographic shift. (　　)
6. There are three main ways in which it could be said that commercial agriculture has contributed to rural emigration. (　　)

Ⅱ. Answer the following questions according to the text.

1. What are the two great movements of population in West Africa?
2. How many major ways in which it could be said that commercial agriculture has contributed to rural emigration?
3. Why the aggregate statistics of regional migration patterns are somewhat elusive?
4. What determines the work conditions?
5. When people are concentrated in central places, what will happen?

Ⅲ. Translate the following terms into their Chinese or English equivalents.

1. affined networks

2. commercial agriculture
3. demographic shift
4. rural social structures
5. agricultural surpluses
6. 农村人口
7. 自然人口增长
8. 物质设施
9. 异族通婚规则
10. 出口农作物农业

Ⅳ. Translate the paragraph into Chinese.

The fact is that the quality and quantity of material amenities are much higher in the cities than in most villages. Moreover, when people are concentrated in central places, markets are bigger and services can be delivered more cheaply on a per capita basis. Village life has its solid virtues, but economic horizons are necessarily limited there, and that limitation matters to a large number of people, especially young adults. The rulers of states, whose own economic behavior draws villagers in thousands to their capital cities, may sometimes regret their presence so close to the corridors of power, where they do more damage if they become upset.

Text B The Traditional Organization of Farming

Any discussion of modern developments in West African agriculture should begin and end with the rural division of labor that constitutes the social context of productive strategies. This chapter begins with a brief recapitulation of traditional economic structure in areas marked either by a complex structure of commodity production or by a simple division of labor. After a detailed examination of forest and savannah agriculture① in the modern period and a more cursory look at the use of livestock, the chapter concludes with an assessment of the effects of these developments on the rural division of labor as a whole.

As we have seen, traditional agriculture was carried out within the framework of a wider division of labor, which was developed to a high degree in some places, notably in the savannah civilizations around the Niger and Senegal rivers. The division of labor was less developed in the intermediate belt and in the forest away from the coast. Nowhere was it more complex than in Hausaland② (northern Nigeria), which contained about a sixth of West

Africa's population in precolonial times. In the advanced societies of the Sudanic zone before the colonial era, agriculture was not the greatest task of all sectors of the population. Many lived in or near towns and cities where they earned their livelihoods principally from manufacturing and services. In places, farming was performed only by slaves or serfs tied to a ruling class whose main business was warfare, slave raiding, and politics. This class derived additional income from taxes on an extraordinary variety of products and activities, which they levied on free men and women who lived in villages partially committed to agriculture. In the rural areas closest to the main centers, farmers were individuated, lacking membership in stable corporate groups and living in production units organized around joint families of agnates. Elsewhere, corporate villages sometimes farmed a common field in order to produce revenues for the state. The range of production strategies was thus rather wide. It was normal in this part of the savannah for an ethnic division of labor to exist between pastoralists and the bulk of agriculturalists. The former were usually called Fulani and they were a major part of the political upheavals that reorganized the region at the beginning of the nineteenth century. Separation between the two groups meant that animals were not used in mixed farming, although intensive manuring was normal in the close-settled zone around Kano and the Voltaic area, for example. Craft specialization reached the point of guilds aimed at urban markets: In the countryside casted specialists were usually endogamous, and they also farmed for themselves, because demand was not high enough to guarantee them a living from crafts alone. Merchants were generally specialized, but in the central areas everyone was in some degree a merchant.

The pervasiveness of commodity economy was variable in the Sudanic zone; transport was so bad that, unless the population was densely settled over a wide area, thirty miles would be sufficient to close off a village's access to high-volume markets. Production of food for the market was thus an option restricted to peoples living near urban centers or to the population of Hausaland in general.

Production of agricultural raw materials for industry (especially cotton for textiles) was well developed in suitable areas. Commercialism was accompanied by a significant emphasis on private property and by the replacement of lineages with cognatic kin groups in which patrilineal descent played a residual part.

Nevertheless, agricultural productivity remained low, although the technology used may have been more labor intensive than that employed elsewhere in West Africa. The basic tool remained the hoe: There was no use made of the plow or irrigation. Animal traction was more or less absent, except for the use of donkeys for transport in some area. Yet even with these limitations on agricultural growth, an extraordinarily varied economy grew up.

Here is M. G. Smith's description of the traditional sexual division of labor in the Hausa province of Zaria during the early 1950s: Men rule, farm, dye, build, work metals, skin, tan and work leather, slay and handle cattle and small livestock, sew all sewn clothes, make musical instruments and music, trade, keep bees, weave mats, may be Mallams (teacher-priests), wash clothes, weave narrow cloth on the men's loom, go on long-distance trading expeditions, make pots, do carpentry—native and European—are the doctors and magicians, the barbers, employed farm laborers, brokers and taxpayers. They also fish, hunt and do all the family marketing, keep goats, sheep, chickens, ducks, turkeys and pigeons, and take part in war.

Women cook for their families, process and sell cooked food (snacks) on their own behalf, sweep and clean the compounds, are solely responsible for delivery and safe care of children, draw wood and water, are custodians of the cult of Bori (spirit possession), tease and spin cotton, weave cloth on the women's broadlooms, thresh, grind and pound corn and food and are, before marriage and when they are old, traders in the markets and from house to house. Women do one another's hair, may (but usually do not) farm, and help with the groundnut, cowpea, cotton and pepper harvests. Very occasionally women become mallams or professional musicians or magicians. In new towns, where prostitutes are numerous, women sometimes work at the dye-pits. Women may not inherit land, nor houses where there are related male heirs, and under Maliki law[3] daughters receive half of the inheritance of sons, but usually inherit the personal possessions of their mothers. Women are expected to provide themselves with luxuries and snacks as they require and on the marriage of daughters, women of the paternal and maternal kin provide most of the dowry. Both men and women make pots and fire, care for chickens, goats and sheep, do hairdressing for their respective sexes, attend the Koranic schools, trade and are required to perform the obligatory duties of Islam.

It should be noted that this list does not include guild or caste specialists such as blacksmiths, jewelers, and the like. Even so, it should serve as a caution to anyone who would subsume traditional agriculture under some such label as "peasant farmers." The Islamic Sudan was and is a sophisticated economy built around a high degree of specialization, of which no element was so deep as the division of labor between the sexes.

In the more remote parts of the savannah and forest, sex, age, and kinship were the sole determinants of the division of labor. The main tasks of small groups formed on the basis of common descent were reproduction, food production, and warfare (defense of territory, women, and movable property); the ideology of kinship relations and divisions according to sex and age generally sufficed to organize these activities. The most important job was to ensure the continuity of the group through a long-term process of recruitment by birth and marriage. Acquisition of wives by a group of men large enough to be demographically viable demanded organization at a level higher than the domestic group. Descent group hierarchies thus gave power and authority to older men over their juniors.

Agriculture was the next most important task. Of all productive activities it occupied the longest time in the year, involved the coordination of the most complex operations, and, at some critical phases, required the cooperation of the largest number of people. In addition, food was the basic source of energy in an economy where clothing and shelter were conveniences rather than necessities and the main use for fuel was cooking. So agriculture dominated production, and other activities were made to fit in with its annual cycle. The other important collective activities—and these united the largest groups—were the common occupation of a territory, defense of it, and ritual regulation of human relations with nature within it. Each of these categories was closely connected to warfare and politics, which in this area were often "tribal"; that is, they consisted of self-help by equally armed groups organized, if at all, by low-level chiefs. But these public activities were essential to the ordinary tasks of production and reproduction and ensured the adherence of small groups to the larger entities that commanded their allegiance.

Notes:

① savannah agriculture 草原农业

② Hausaland 豪萨兰（尼日利亚北部）

③ Maliki law 马利基法律

Exercises

Work in groups and discuss the following questions.

1. What is the main idea of this passage?
2. Why, in the advanced societies of the Sudanic zone before the colonial era, wasn't agriculture the greatest task of all sectors of the population?
3. What are the limitations on agricultural growth in West Africa?
4. What were the sole determinants of the division of labor in the more remote parts of the savannah and forest?
5. What were Men's roles according to M. G. Smith's description of the traditional sexual division of labor in the Hausa province of Zaria during the early 1950s?

Unit 7 Plants and Animals

Text A Early Human Societies and Their Plants

The development of agriculture is universally regarded as one of the defining moments in the evolution of humankind. Indeed, many accounts of human development still describe the so-called "invention" of agriculture as if it were a sudden and singular transformative event.

The acquisition of the know-how and technology that enabled people to practice agriculture is conventionally portrayed as a dramatic and revolutionary change, which occurred about 11,000 years ago at the start of the Neolithic period① (or "New Stone Age").

We are told that this revolutionary event completely altered the diet, lifestyle, and structure of the human societies involved, most notably in the Near East. The epochal "invention" of agriculture is then supposed to have led directly to urbanization and quantum leaps in technological and artistic development as part of a unidirectional and profoundly progressive process. This notion of a sudden agricultural revolution originated because of what appeared to be the almost overnight appearance and cultivation of new forms of several key plants, especially cereals and pulses, that had supposedly been deliberately "domesticated" by people. Almost simultaneously, so it seemed, the new farming-based cultures began to build increasingly complex, permanent habitations that soon developed into elaborate urbanized cultures and, eventually, civilizations with imperial aspirations. Moreover, it was also originally believed, and is still repeated in a surprisingly large number of textbooks, that agriculture was somehow "invented" in the Near East and subsequently exported to Europe, Africa, and the Far East. The entire process of agricultural and societal development has also been decorated with Enlightenment② and Victorian overtones of inevitability and progression, as if humanity was somehow "destined" to tame plants and animals and to develop complex, technologically based societies. This "revolutionary" thesis of the origins of agriculture is now being successfully challenged by manifold lines of evidence from a spectrum of scientific disciplines that includes archaeology, geology, climatology, genetics, and ecology.

It is now clear that several human cultures (possibly numbered in the dozens) independently developed distinctive systems of agriculture on at least four different continents.

Over the past decade or so, detailed archaeological and genetic evidence has emerged supporting the view that widespread cultivation of crops evolved separately in various parts of Asia, Africa, Mesoamerica③, and South America.

In contrast, in Europe, North America, and Australasia, crop cultivation occurred much later. In these latter three regions, crops and agronomic techniques were only secondarily acquired from the primary agricultural societies. These crops were then grown in places that were far from their initial centres of origin. In the comparatively few primary centres of crop cultivation, a relatively narrow range of locally available edible plants was domesticated as the major food staples. Wherever suitable species were available, it was the large-grained cereals that were the most favoured candidates for cultivation as staple crops. The most obvious examples are rice, wheat, and maize; these three plants were among the earliest domesticates and are still by far the most important crops grown across the world, supplying well over two-thirds of human calorific needs. The second most popular class of staple domesticants were the starchy tubers such as yams and potatoes, but these crops were not as versatile as cereals, especially as regards long-term storage, and this limited their more general use. The major class of supplementary crop is the pulses, or edible seeded legumes, which provide useful proteins and nutrients lacking in cereals and tubers, as well as replenishing soil fertility with nitrogen compounds. Domestication of these different crop species did not occur at the same time or in the same place.

Several overlapping, and sometimes lengthy, primary domestication processes were in progress around the world over a period of at least eight millennia from about 13,000 BP until 5,000 BP. In several cases, such as wheat and rice, a single plant species was domesticated completely independently on numerous occasions, by various unrelated human cultures living in different periods and in different regions of a continent. Moreover, it now appears that the systematic cultivation of crops was preceded in most places by an extremely lengthy preagricultural phase④ of plant husbandry. During this period, many geographically unconnected groups of humans started to collect, process, and even manage certain favoured plants for food use, while still relying on a nomadic hunter-gathering lifestyle to sustain the bulk of their livelihoods. In the Near East, this prefarming phase⑤ of informal plant management may have extended for many millennia and perhaps tens of millennia, from as long ago as 40,000 or 50,000 BP. It is also important to realize that agriculture is by no means the only successful and enduring option for the management and exploitation of plants. Indeed, numerous societies around the world opted over many millennia to remain wedded to a more flexible lifestyle of informal nurturing and collection of wild plants, rather than committing themselves to full-time agriculture.

Notes:

① Neolithic period 新石器时代

② Enlightenment 启蒙运动

③ Mesoamerica 中美洲

④ preagricultural phase 前农业阶段

⑤ prefarming phase 前种植阶段

Exercises

Ⅰ. Read each of the following statements carefully and decide whether it is true or false according to the text.

1. In Europe, North America, and Australasia, crop cultivation occurred much earlier. ( )
2. The epochal "invention" of agriculture is then supposed to have led directly to urbanization and quantum leaps in technological and artistic development as part of a unidirectional and profoundly progressive process. ( )
3. The development of agriculture is universally regarded as one of the defining moments in the evolution of humankind. ( )
4. It is also important to realize that agriculture is the only successful and enduring option for the management and exploitation of plants. ( )
5. It is now clear that several human cultures (possibly numbered in the dozens) independently developed distinctive systems of agriculture on at least four different continents. ( )
6. There is sufficient evidence to prove that a single plant species, such as wheat and rice, was domesticated with joint efforts of several cultures. ( )

Ⅱ. Answer the following questions according to the text.

1. Why was there a notion of a sudden agricultural revolution?

2. Is it right to say that agriculture was somehow "invented" in the Near East and subsequently exported to Europe, Africa, and the Far East?
3. What did detailed archaeological and genetic evidence tell us?
4. Why did crop cultivation occur much later in Europe, North America, and Australasia?
5. What were the second most popular class of staple domesticants?

Ⅲ. Translate the following terms into their Chinese or English equivalents.

1. agricultural revolution
2. exploitation of plants
3. archaeological and genetic evidence
4. distinctive systems of agriculture
5. prefarming phase
6. 人类文化
7. 城市化
8. 可食用植物
9. 主要农作物
10. 决定性时刻

Ⅳ. Translate the paragraph into Chinese.

Over the past decade or so, detailed archaeological and genetic evidence has emerged supporting the view that widespread cultivation of crops evolved separately in various parts of Asia, Africa, Mesoamerica, and South America. In contrast, in Europe, North America, and Australasia, crop cultivation occurred much later. In these latter three regions, crops and agronomic techniques were only secondarily acquired from the primary agricultural societies. These crops were then grown in places that were far from their initial centres of origin.

Text B The Traveling Exotic Animal Protection Act

Could Ban Circus Animals

What is the Traveling Exotic Animal Protection Act? The Traveling Exotic Animal Protection Act, a. k. a. TEAPA, would restrict the use of exotic animals in traveling circuses in the United States. The bill is sponsored by Rep. Jim Moran (D, VA), who worked with Animal Defenders International (ADI)[①], and the Performing Animal Welfare Society (PAWS)[②], and was announced at a press conference on November 2, 2011 with celebrities

Bob Barker[3] and Jorja Fox[4].

Moran is the co-chair of the Congressional Animal Protection Caucus. According to Moran's press release, "The 'Traveling Exotic Animal Protection Act' would comprehensively tackle the use of all exotic animals in circuses. The bill would end the keeping of animals for extended periods in temporary facilities, and the cruel training and control methods employed by circuses and address public safety issues."

In 2009, Bolivia enacted a ban on circus animals[5] that took effect in 2010. In January of 2011, ADI rescued lions, monkeys, a baboon and other animals seized from Bolivian circuses. In 2010, China became the second nation to ban circuses, with Peru following in 2011.

What is prohibited? According to a bill summary distributed by ADI, "TEAPA Amends S. 13 of the Animal Welfare Act to the effect that no exhibitor may allow the (use) of an exotic or wild animal (including a non-human primate) in an animal act if it is living in mobile accommodation and is constantly traveling (the United States)." If passed, this bill would apply to elephants, lions, tigers, monkeys, bears and other animals.

What are the exceptions? The bill does not apply to zoos, aquariums, universities, laboratories, rodeos, horse racetracks or facilities where the animals reside permanently and are not constantly traveling. The bill also does not apply to domestic animals.

Why ban exotic animals? From cruel training techniques to the misery of confinement, there are many abuses that animals in circuses suffer. While animals in circuses are covered by the Animal Welfare Act, the AWA does not prohibit the use of bullhooks, whips, electrical shock prods, or other such training devices. Elephants, because they are so large and wander for miles every day in the wild, suffer immensely in circuses.

ADI also points out that exotic animals are a threat to public safety, citing numerous incidents in recent years, such as: April 9, 2010/Wilkes-Barre, Pennsylvania: An animal handler with the Hamid Circus was kicked and thrown about 20 feet by an African elephant named Dumbo at Irem Shrine Circus. The handler died at the scene from multiple traumatic injuries.

On November 4, 2009, an elephant escaped from the Family Fun Circus in Enid, Oklahoma and was struck by an SUV on US Interstate 81.

March 7, 2009/Indianapolis, Indiana: At least 15 children and one adult were injured when an elephant who was being used to give rides at the Murat Shrine Circus became startled, stumbling and knocking over the scaffolding stairway leading to the elephant ride.

These are compelling reasons to ban the use of exotic animals in circuses, but from an animal rights viewpoint, using any animal— domestic or wild—violates that animal's right to live free of human exploitation. Even if negative reinforcement were not used, merely confining the animals and forcing them to do tricks is exploitative.

What you can do: Urge your federal legislators to support the Traveling Exotic Animal Protection Act. You can look up your representative on the House of Representatives website, while your senators can be found on the official Senate website. As of November 3, 2011, the bill did not have a bill number yet, but can be identified as the "Traveling Exotic Animal Protection Act."

Notes:

① Animal Defenders International (ADI) 国际动物保护组织

② Performing Animal Welfare Society (PAWS) 表演动物福利协会

③ Bob Barker 鲍勃·巴克(出生于1923年12月12日),美国前电视游戏节目主持人。巴克长期以来一直是动物权利和动物权利运动的支持者,包括动物权利联合活动家和海洋守护者协会等团体

④ Jorja Fox 乔雅·福克斯(出生于1968年7月7日),美国女演员、音乐家和词曲作家。福克斯从19岁起就一直是一名虔诚的素食主义者,与善待动物组织合作,帮助推广素食主义,并与ADI合作,为马戏团动物的苦难带来光明

⑤ circus animal 马戏团动物

Exercises

Work in groups and discuss the following questions.

1. What is the Traveling Exotic Animal Protection Act?
2. What is prohibited in this Act?
3. What are the exceptions in this Act?
4. Why ban exotic animals?
5. What can you do according to the text?

Unit 8 Computer Science

Text A Supplemental Skills for Success in 3D

If you aspire to success as a 3D artist, the best way to improve is to practice as much as possible. Like everything in life, you're only limited by the amount of time and effort you're willing to put in. But it's also important to add some variety to your daily routine. If you spend every single day "modeling to infinity" (as I've heard it affectionately called) it's easy to burn yourself out.

There are *a lot* of things you can be doing when you're not modeling or animating that will help you improve nonetheless. In this two part article, we provide suggestions for non-3D skills that can be beneficial to anyone working in the computer graphics industry:

Life Drawing①

Here's a conversation that occurs pretty much every day at one of the popular 3D forums on the web: "Hey guys, I'm an aspiring 3D artist, and I'm just wondering—is it possible to become a modeler without being good at drawing?" "Yeah—but you should really start life drawing."

Life drawing is often seen as the "granddaddy" of all skills in the artistic world, because if you're truly great at it you're pretty much worth your weight in gold. I don't think I need to go through the long list of opportunities that present themselves when you're a master figurative artist.

In terms of 3D—yes, it's possible to make it in the 3D world without knowing how to draw well, but if you want an advantage in a competitive industry this is a great place to start. The better you are at life drawing the better you'll be at designing characters, because instead of being bogged down by anatomy & proportion, you'll be free to focus on the thing that matters most—design.

Like anything, it takes long hours, repetition, and incredible discipline to get really good at figure drawing, but the payoff is very, very valuable.

George Bridgman and Andrew Loomis are the undisputed masters when it comes to teaching life drawing. Here are their seminal texts, if you're feeling bookish. Loomis is pricey because the paperbacks are out of print, but Bridgman can be had for around ten bucks. I used to carry a copy *everywhere*:

Figure Drawing For All It's Worth(Andrew Loomis)

Bridgman's Complete Guide to Drawing from Life

Color Theory

Nothing messes with an image like a bad color scheme. Color theory is one of those things they try to teach you in high-school art class, but the reality of the subject is far deeper than color wheels and Andy Warhol② slides.

Some people seem to have an inherent grasp on color. Others (like me) need a little help.

James Gurney, who wrote and illustrated the *Dinotopia*③ series, released the book *Color and Light* late last year, and after 60 reviews it still carries a 5 star rating on Amazon. It's the most eye-opening resource on practical color theory I've ever come across, and I honestly can't recommend it enough. I wish I'd had access to it ten years ago.

Figurative Sculpture④

If you're looking to become a character artist, it's not a bad idea to brush up on your traditional sculpture from time to time. Figurative sculpture, like life drawing, will help you learn anatomy inside and out so that when you're modeling you can focus on design and originality.

Most sculpting workflows require the artist to focus on large forms first and details later. With clay or polymer there's a lot less temptation to preemptively detail, knowing you don't have an undo button to rely on. This is one of the best habits you can acquire as an artist, and it will only make you a more efficient modeler.

Believe it or not, there's still a high prevalence of traditional maquette building⑤ in the film industry, so figurative sculpture could actually be a backdoor into the industry if you get good enough at it.

Photography

Ask any professional photographer—it's all about the light. As it would happen, light is pretty important in the computer graphics world too—the best model in the world will fail to produce a good render if you don't have a good lighting solution.

If you're an aspiring 3D artist, or if you see lighting as a weak point in your workflow, try to pick up a camera and do some photography. After some time behind a lens you'll start to learn what works and what doesn't, and what sort of lighting produces compelling imagery.

Another advantage of picking up photography (or studying good photographs)—it's probably the fastest and easiest way to learn composition.

Acting

Animators are actors. Simple as that. If you really want to become a great animator, do yourself a favor and act! Take a class. Join an improv group. Find a friend and do scenes with them.

Or, if you're shy, start to film yourself acting out the scenes you want to animate. Nobody has to see them, but it'll provide great reference material for you to animate from, and seeing yourself on film makes it easier to spot the things that don't work.

Architecture

Architecture is absolutely central to the life of an environment artist. There's a saying in the 3D industry that "form should always follow function." What's meant by this is, let your imagination run wild, but always try to base your designs in reality.

I'm not saying you need to go out and study the engineering side of architecture, but having some knowledge of architectural history can never hurt. You should *always* have an architectural reference on hand when you're designing for a specific culture/period. Contrary to what you may have heard, there are no bonus points for designing "from imagination."

Notes:

① life drawing 人体素描

② Andy Warhol 安迪·沃霍尔(1928年8月6日—1987年2月22日),20世纪艺术界最有名的人物之一,波普艺术的倡导者和领袖

③ *Dinotopia* 恐龙帝国(电视剧名)

④ figurative sculpture 具象雕塑

⑤ maquette building 初步设计的模型构建

Exercises

Ⅰ. Read each of the following statements carefully and decide whether it is true or false according to the text.

1. If you aspire to success as a 3D artist, the best way to improve is to practice as much as possible. (　　)
2. Life drawing is often seen as the "father" of all skills in the artistic world. (　　)
3. The better you are at life drawing the better you'll be at designing characters. (　　)
4. Most sculpting workflows require the artist to focus on details first and large forms later. (　　)
5. Architecture is absolutely central to the life of an environment artist. (　　)
6. You should always have an architectural reference on hand when you're designing for a specific culture/period. (　　)

Ⅱ. Answer the following questions according to the text.

1. How to be a successful 3D artist?
2. Why life drawing is seen as the "granddaddy" of all skills in the artistic world?
3. Who wrote and illustrated the *Dinotopia* series?
4. What are the advantages of picking up photography?
5. If you really want to become a great animator, what should you do?

Ⅲ. Translate the following terms into their Chinese or English equivalents.

1. life drawing
2. competitive industry
3. undisputed masters
4. color wheels
5. inherent grasp
6. 具象雕塑
7. 角色艺术家
8. 雕刻的工作流
9. 模型构建
10. 环境艺术家

Ⅳ. Translate the paragraph into Chinese.

Ask any professional photographer—it's all about the light. As it would happen, light is pretty important in the computer graphics world too—the best model in the world will fail to produce a good render if you don't have a good lighting solution. If you're an aspiring 3D artist, or if you see lighting as a weak point in your workflow, try to pick up a camera and doing some photography. After some time behind a lens you'll start to learn what works and what doesn't, and what sort of lighting produces compelling imagery.

Text B Cyberterrorism: Latest Threat to National Computer Security?

Cyberterrorism①: Invading Cyberspace and Networks as a Form of Terrorist Attack. The immediate aftermath of 9/11 produced several interrelated fears about the role that the Internet and networked computers might play in a terrorist attack. First, the government and the private sector worried that information housed on computers might be vulnerable. Those who seek to threaten American national security could do so if they obtained weapons system information or substantial financial information from banks. A second, related concern was that the networks over which this information is transmitted could be compromised.

Their compromise would of course give the criminally-minded among us access to information, as well as the opportunity to influence or manipulate its flow.

Defining cyberterrorism has been a challenge. Cyberterrorism is not necessarily designed to cause a terrifying visual spectacle that can be exploited for propaganda purposes, as conventional terrorism is.

Defining cyberterrorism as broadly as possible serves those who want to expand control over cyberspace. Even though only a small amount of cybercrime could actually be designated terrorism, the term "cyberterrorism" creates a subliminal linkage in listener's minds to groups such as Al Qaeda② and other global jihadists③. The government and private industry can in turn use fears about Al Qaeda to create support for tighter controls on electronic information.

Cyberterrorism, Is it a Threat? A number of experts conclude that while the threat of cyberterrorism does exist, it is routinely exaggerated by various actors for political or financial gain. As they note, the United States (and indeed, much of the rest of the world) is profoundly dependent on computer networks for its daily well functioning.

It is important to safeguard critical infrastructure from vulnerability to attack.

At the same time, maintaining a state of anxiety over that vulnerability can be profitable. As Gabriel Weimann pointed out in Cyberterrorism: How Real is the Threat? ... an entire industry has emerged to grapple with the threat of cyberterrorism ... private companies have hastily deployed security consultants and software designed to protect public and private targets ... the federal government requested $ 4.5 billion for infrastructure security [following 9/11], and the FBI now boasts more than one thousand "cyber investigators."

US National Computer Security Efforts

The Patriot Act addresses cyberterrorism: The Bush Administration has offered to counter the threat of cyberterrorism by expanding the punishments for those committing cybercrimes, and pushing for Internet Service Providers to disclose information to the government for investigations. Both of these provisions appear in the US Patriot Act (the 2001 Act was updated in 2005). According to a White House press release, the act also usefully "allows Internet providers, without fear of being sued, to give information to law enforcement when it would help law enforcement prevent a threat of death or serious injury." However, a September, 2007 ruling by Manhattan federal judge Victor Marrero on the constitutionality of National Security Letters (NSLs) Challenges the Patriot Act provision.

The Patriot Act allowed the government to issue NSLs requesting customer information without legal approval to a company, and also put "gag orders" on companies over their ability to disclose their receipt of these letters. Following an ACLU lawsuit brought on behalf of an Internet company (which remains anonymous), Judge Marrero ruled the gag order violates the first amendment right to free speech, and the lack of judicial oversight violates the fourth amendment prohibition of "unreasonable search and seizure."

New offices address cyberterrorism: The Department of Homeland Security houses a National Cyber Security Division which aims to build a national cyberspace response system, and to put into use a cyber-risk management program to protect critical infrastructure.

Additionally, the Air Force has established a provisional Cyber Command, which will coordinate with air and space operations to conduct wars in cyberspace.

Notes:

① cyberterrorism 网络恐怖主义

② Al Qaeda 基地组织(著名恐怖主义组织领袖为拉登)

③ jihadists 圣战者;伊斯兰圣战者

Exercises

Work in groups and discuss the following questions.

1. What is cyberterrorism?
2. Is cyberterrorism a threat?
3. What are the Computer Security Efforts made by US?
4. What is the function of the Patriot Act?
5. What is the aim of the National Cyber Security Division?

Unit 9 Information and Communication Technology

Text A An Introduction to Information Technology (IT)

The terms "information technology" and "IT" are widely used in business and the field of computing. People use the terms generically when referring to various kinds of computer-related work, which sometimes confuses their meaning.

What Is Information Technology?

A 1958 article in *Harvard Business Review*① referred to information technology as consisting of three basic parts: computational data processing, decision support, and business software.

This time period marked the beginning of IT as an officially defined area of business; in fact, this article probably coined the term.

Over the ensuring decades, many corporations created so-called "IT departments" to manage the computer technologies related to their business. Whatever these departments worked on became the *de facto*② definition of Information Technology, one that has evolved over time.

Today, IT departments have responsibility in areas like:

- computer tech support
- business computer network and database administration
- business software deployment
- information security

Especially during the dot-com boom of the 1990s, Information Technology also became associated with aspects of computing beyond those owned by IT departments. This broader definition of IT includes areas like:

- software development
- computer systems architecture
- project management

Information Technology Jobs and Careers

Job posting sites commonly use IT as a category in their databases. The category includes a wide range of jobs across architecture, engineering and administration functions. People with jobs in these areas typically have college degrees in computer science and/or information systems.

They may also possess related industry certifications. Short courses in IT basics can also be found online and are especially useful for those who want to get some exposure to the field before committing to it as a career.

A career in Information Technology can involve working in or leading IT departments, product development teams, or research groups.

Having success in this job field requires a combination of both technical and business skills.

Issues and Challenges in Information Technology

1. As computing systems and capabilities continue expanding worldwide, *data overload* has become an increasingly critical issue for many IT professionals. Efficiently processing huge amounts of data to produce useful business intelligence requires large amounts of processing power, sophisticated software, and human analytic skills.

2. *Teamwork and communication* skills have also become essential in most businesses, to manage the complexity of IT systems. Many IT professionals are responsible for providing service to business users who are not trained in computer networking and other information technologies, but who are instead interested in simply using the technology to get their work done efficiently.

3. *System and network security* issues are a primary concern for many business executives, as any security incident can potentially damage a company's reputation and cost large sums of money.

Computer Networking and Information Technology

Because networks play a central role in the operation of many companies, business computer networking topics tend to be closely associated with Information Technology.

Networking trends that play a key role in IT include:

- *Network capacity and performance*: The popularity of online video has greatly increased the demand for network bandwidth both on the Internet and on IT networks. New types of software applications that support richer graphics and deeper interaction with computers also tend to generate larger amounts of data and hence network traffic. Information technology teams must plan appropriately not just for their company's current needs but also this future growth.
- *Mobile and wireless usages*: IT network administrators must now support a wide array of Smartphone and tablets in addition to traditional PCs and workstations. IT environments tend to require high-performance wireless hotspots③ with roaming capability④. In larger office buildings, deployments are carefully planned and tested to eliminate dead spots⑤ and signal interference.
- *Cloud services*: Whereas IT shops in the past maintained their own server farms⑥ for hosting email and business databases, some have begun migrating to cloud computing environments, where third-party hosting providers maintain the data. This change in computing model dramatically changes the patterns of traffic on a company network but also requires significant effort in training employees on this new breed of applications.

Notes:

① *Harvard Business Review*《哈佛商业评论》

② *de facto* "de facto"是一个拉丁语表达,意思是"关于事实"。在法律上,它通常意味着"在实践中但不一定是法律规定的"或"在实践或现实中,但不是正式确立的"。

③ wireless hotspot 无线热点

④ roaming capability 漫游功能

⑤ dead spot 哑点;死点;死角;布线死区

⑥ server farm 服务器群组;服务器群;服务器农场

Exercises

Ⅰ. Read each of the following statements carefully and decide whether it is true or false according to the text.

1. Over the ensuring decades, many corporations created so-called "IT departments" to manage the computer technologies related to their business. ( )
2. Job posting sites commonly use IT as a category in their databases. ( )

3. Short courses in IT basics can only be found online and are especially useful for those who want to get some exposure to the field before committing to it as a career. ( )
4. System and network security issues are not primary concerns for many business executives. ( )
5. The popularity of online video has greatly increased the demand for network bandwidth both on the Internet and on IT networks. ( )
6. Information technology teams must plan appropriately just for their company's current needs. ( )

Ⅱ. Answer the following questions according to the text.

1. What is Information Technology?
2. What are the challenges in Information Technology?
3. What are the responsibilities of IT departments?
4. Why do business computer networking topics tend to be closely associated with Information Technology?
5. What are the requirements for having success in IT job field?

Ⅲ. Translate the following terms into their Chinese or English equivalents.

1. information technology
2. computational data processing
3. database administration
4. business software deployment
5. information security
6. 软件开发
7. 计算机系统架构
8. 项目管理
9. 云服务
10. 计算模型

Ⅳ. Translate the paragraph into Chinese.

As computing systems and capabilities continue expanding worldwide, *data overload* has become an increasingly critical issue for many IT professionals. Efficiently processing huge amounts of data to produce useful business intelligence requires large amounts of processing power, sophisticated software, and human analytic skills.

Text B Parent Connection Goals Are Best Served by Technology

While connecting and communicating with parents has always meant engaging in regular, two-way, meaningful dialogue, the major difference today is the methodology used for that parent-teacher interaction. Technology is emerging as a critical component in parent-teacher communication.

Today, social media and school websites help to create a presence in a local community for all stakeholders. In addition, there has been an increase in the kinds of software programs, from quarterly grade reports to daily homework reminders that can used by parents and teachers alike in order to stay informed about student learning.

Moreover, connecting with parents and the community is a goal written into many teacher evaluation programs. In these programs, there is a digital record of the kinds of communication that teachers and parents have used.

Reviewing what has been done in the past, with or without a digital record, can be a starting point in determining what kinds of technologies are best suited for connecting and communicating with parents, particularly in a secondary school when efforts may need to be coordinated.

Assess What Technology Is Used for Communication

The National Parent-Teacher Association (PTA)① Standards for Family-School Partnerships suggests that "families, the community, and school staff communicate in numerous interactive ways, both formally and informally". A first step assessing is to analyze what kinds of programs or apps are currently being used by all parties and what information is best suited for each of these technologies. For example, websites are best for static information that does not change, while social media platforms (ex: Twitter, Facebook, Instagram, Pintrest, etc.) are generally better suited for publicizing specific events.

Here are ten (10) questions, several adapted from the National PTA's website, to help in analyzing the use of technology:

1. When is a technology best suited for general communication between parents and

teachers?

2. What is best platform for widespread news/message dissemination to all stakeholders?
3. Which software/apps are best suited for private communication about a specific student?
4. What technologies are already being used effectively by both parents and teachers?
5. What types of hardware do parents and teachers have access to on a regular basis? (computers, smartphones, land-lines, etc.)
6. What kinds of skills do parents have with technology and to what extent are they using it?
7. What kinds of skills do educators have with technology and to what extent are they currently using those skills?
8. How can face-to-face communication between parents and educators be supported (not replaced) with technology?
9. How can teachers in different disciplines (Gr 7－12) collaborate and coordinate separate communication using technology?
10. Who is the "gatekeeper" for outgoing communication in a 7–12 school?

Additional Reasons to Communicate Using Technology

Time is the best reason for educators to design communication using technology. Websites and grade portals② allow parents access to information 24/7. Posted information can be regular and made timely with updates. For those parents on less traditional work schedules or for those parents who cannot physically travel to the school face-to-face meetings can be organized on digital video platforms.

For educators, schools now have databases for e-mail addresses, for robo-calls and for text messages in order to make parent contact. These technologies are far more efficient in the speed of delivery for regular reminders of for emergencies.

Language barriers may have been a problem in the past, but there is multiple (free) translation apps can help facilitate communication with parents or stakeholders③ who do not speak English. Information can be posted and updated in multiple languages using these applications.

Equal access to information is increased with technology. While hardware for technology can be expensive, access to information is an "equalizer" in education. As they do with students, educators can communicate to parents the same access, regardless of income or ethnicity, to a wide set of resources through technology.

In addition, students in grades 7–12 are generally more adept at using technology and as stakeholders in their own education can share with their parents the responsibilities for communication on digital platforms.

Whether it is a platform for one-way information posting or one that facilitates a two-way dialogue, technology is emerging as a critical component in increasing parent-teacher communication.

Notes:

① The National Parent-Teacher Association (PTA) 全国家长教师联谊会

② grade portal 年级门户

③ stakeholder 利益相关者;利害关系人

Exercises

Work in groups and discuss the following questions.

1. What are the suggestions given by the National Parent-Teacher Association (PTA) Standards for Family-School Partnerships?
2. What are the ten questions?
3. What is the best reason for educators to design communication using technology?
4. Why do schools now have databases for e-mail addresses, for robo-calls and for text messages?
5. What are the additional reasons to communicate using technology?

Unit 10 Energy Science

Text A All Types of Coal Are Not Created Equal

Coal is a sedimentary black or dark brown rock that varies in composition. Some types of coal burn hotter and cleaner while others contain high moisture content and compound that contribute to acid rain and other pollution when they're burned.

Coals of varying composition are used as a combustible fossil fuel① for generating electricity and producing steel around the world. It has been the fastest growing energy source worldwide in the 21st century, according to the International Energy Agency②.

People don't "produce" coal. Geological processes and decaying organic matter create it over thousands of years. It's mined from underground formations or "seams," through underground tunnels, or by removing large areas of the earth's surface. The excavated coal must be cleaned, washed, and processed to prepare it for commercial use.

China currently produces more coal than any other country in the world, although its proven reserves rank fourth behind the US, Russia, and India. The IEA estimates that global supply should increase at a rate of about 0.6 percent through 2020.

Australia tops the worldwide list of exporters, having sent 298 million metric tons of coal overseas in 2010. Indonesia and Russia ranked second and third, exporting 162 and 109 million metric tons respectively. The US came in fourth globally, having shipped 74 million metric tons beyond its borders that same year.

South Africa relies most heavily on coal, taking 93 percent of its electric power from this energy source.

China and India also rely heavily on coal for substantial amounts of their energy at 79 and 69 percent respectively. The US takes 45 percent of its electricity from this source, ranking it

11th on the global list of countries that generate power from this source.

Coal falls into two main categories: hard and soft. Soft coal is also known as brown coal or lignite. China produces more hard coal than any other country by a factor of about three. The whopping 3,162 million metric tons of hard coal produced by China dwarfs the output of the second and third ranked producers—the US at 932 million metric tons and India at 538 million metric tons.

Germany and Indonesia nearly tie for the honor of top honors in the production of soft brown coal. These countries dug up 169 million and 163 million metric tons respectively.

Coking coal, also known as metallurgical coal, has low sulfur and phosphorus content and is able to withstand high heat. Coking coal is fed into ovens and subjected to oxygen-free pyrolysis, a process that heats the coal to approximately 1,100 degrees Celsius. This melts it and drives off any volatile compounds and impurities to leave pure carbon. The hot, purified, liquefied carbon solidifies into lumps called "coke" that can be fed into a blast furnace along with iron ore and limestone to produce steel.

Steam coal③, also known as thermal coal, is suitable for electric power production. Steam coal is ground into a fine powder that burns quickly at high heats and is used in power plants to heat water in boilers that run steam turbines. It may also be used to provide space heating for homes and businesses.

All types of coal contain fixed carbon, which provides stored energy and varying amounts of moisture, ash, volatile matter, mercury and sulfur. Because the physical properties and coal quality vary widely, coal-fired power plants must be engineered to accommodate the specific properties of available feedstock and to reduce emissions of pollutants such as sulfur, mercury and dioxins.

The stored energy potential within coal is described as the "calorific value", "heating value" or "heat content." It's measured in Btu or MJ/kg. Btu stands for British thermal unit④, the amount of heat that will warm approximately 0.12 US gallons—a pound of water — by one degree Fahrenheit at sea level. Btu is sometimes written as BTU.

MJ/kg stands for millijoule per kilogram and is the amount of energy stored in a kilogram. This is an expression of energy density for fuels measured by weight.

Coal releases thermal energy or heat when it is burned, along with carbon and ash. Ash is made up of minerals such as iron, aluminum, limestone, clay and silica, as well as trace elements such as arsenic and chromium.

The international standards organization ASTM⑤ has issued a ranking method for classifying grades of coal formed from biodegraded peat-based humic substances and organic material or vitrinite. The coal ranking is based on levels of geological metamorphosis, fixed carbon, and calorific value. It is known as ASTM D388-05 Standard Classification of Coals by Rank.

How do the four types compare? As a general rule, the harder the coal, the higher its energy value and rank. The following is a comparative ranking of four different types of coal from the most dense in carbon and energy to the least dense:

Rank	Type of Coal	Calorific Value (MJ/kg)
1	Anthracite	30
2	Bituminous	18.8–29.3
3	Sub-bituminous	8.3–25
4	Lignite (brown coal)	5.5–14.3

Notes:

① combustible fossil fuel 燃烧化石燃料

② International Energy Agency (IEA) 国际能源机构

③ steam coal 动力煤；锅炉煤

④ British thermal unit(Btu) 英国热量单位

⑤ ASTM (American Society for Testing and Materials) 美国材料与试验协会

Exercises

Ⅰ. Read each of the following statements carefully and decide whether it is true or false according to the text.

1. Some types of coal burn hotter and cleaner while others contain high moisture content and compound that contribute to acid rain and other pollution when they're burned. (　　)
2. China currently produces more coal than any other country in the world. (　　)
3. The IEA estimates that global supply should increase at a rate of about 0.8 percent through 2020. (　　)
4. Russia and Indonesia ranked second and third, exporting 162 and 109 million metric tons respectively. (　　)
5. The US came in fourth globally, having shipped 74 million metric tons beyond its borders that same year. (　　)
6. South Africa relies most heavily on coal, taking 93 percent of its electric power from this

energy source. (　　)

Ⅱ. **Answer the following questions according to the text.**

1. Where does coal come from?
2. Which country tops the worldwide list of exporters?
3. How many main categories of coal?
4. What is "calorific value"?
5. What is the standard for coal ranking?

Ⅲ. **Translate the following terms into their Chinese or English equivalents.**

1. combustible fossil fuel
2. geological processes
3. coking coal
4. steam coal
5. brown coal
6. 挥发性化合物
7. 杂质
8. 铁矿石
9. 物理性质
10. 热量单位

Ⅳ. **Translate the paragraph into Chinese.**

Coals of varying composition are used as a combustible fossil fuel for generating electricity and producing steel around the world. It has been the fastest growing energy source worldwide in the 21st century, according to the International Energy Agency. People don't "produce" coal. Geological processes and decaying organic matter create it over thousands of years. It's mined from underground formations or "seams," through underground tunnels, or by removing large areas of the earth's surface. The excavated coal must be cleaned, washed, and processed to prepare it for commercial use.

Text B　Energy for Future Presidents

Richard A. Muller① made a splash when, through a "conservatively-funded" research project, his team returned results that strongly confirmed the scientific consensus in support of man-made global warming due to carbon dioxide (and other greenhouse gas) emissions. As a follow up to those headlines, and also his previous book *Physics for Future Presidents*, he dives deeply into all aspects of the energy debate, explaining it in terms that are both technically precise but also easily accessible to the average reader.

While some of the details may cause the non-science-enthusiasts eyes to gloss over, most of the specific numbers can be skimmed over to reach the more accessible explanations.

Muller divides his book into 5 distinct sections, which each focuses on different elements of the current energy debate in our country.

He is a scientist, so the book is mostly told from that perspective, giving an overview of the science and options rather than prescribing a definite strategy, though the last section does narrow the perspective on his specific recommendations about how to proceed.

Energy Catastrophes Muller begins his book by addressing the energy related disasters that have often dominated the media discussions around energy policy. He has a chapter on each of the following three energy catastrophes, though his overall conclusions are that the first two are not as catastrophic as the public has been led to believe, though the third is very real and requires a prominent position in guiding energy policy:

Oddly, though he has a chapter devoted to global warming, he does not have a chapter in this section devoted to energy security (though he does have a 5-page section on the subject in Part 2), despite the fact that he puts energy security alongside global warming as one of the driving factors in the energy debate. Presumably this is because energy security isn't so much an energy catastrophe as a *pending* energy catastrophe.

The Energy Landscape This section has 4 chapters focusing on the important sources of power production with current energy policy and technologies.

- *The Natural Gas Windfall* America has access to a lot of natural gas, especially now that the process known as "fracking" has shown a cost-effective way to access natural gas that had previously been unreachable. Though Muller clearly supports natural gas as having a prominent role within America's energy solution, he does not shy away from the science controversies and questions related to fracking, encouraging further research in this area.
- *Liquid Energy Security* This chapter is critically important to Muller's overall discussion of energy policy. Muller doesn't believe that the US has an energy crisis, but rather a crisis in easily accessing liquid energy without being dependent upon other nations... and that's where the energy security issue comes in.

- *Shale Oil*② As with fracking, Muller's stance on this is to explain as much detail as he can about the process, clearly identifying the controversies and questions related to shale oil extraction and urging further study of this new technology.
- *Energy Productivity* The best way forward with energy policy is to work on ways that can reduce the need for energy, which results in a bigger benefit (both environmental and economic) for most key players. Here he outlines some current policy approaches that help improve energy productivity, such as the "decoupling plus" process used in California.

1. **Alternative Energy** This section focuses on various forms of energy which Muller classifies as alternatives to our most traditional (and environmentally dirty) energy sources: gasoline, coal, and even natural gas.
2. **What is Energy**? This section is pure physics, focusing on the scientific meaning of energy, including some basic thermodynamics concepts about the flow of energy.
3. **Advice for Future Presidents** Here Muller goes a bit beyond his merely informative position to offer his personal recommendations about energy policy, based upon the information presented in the rest of the book. Don't worry, though, because he isn't particularly long-winded about it. This whole section is less than 15 pages.

If you're looking for a book where you'll be gripped by every page, this book is not for you. Education, not entertainment, is the primary objective. He does, for example, include equations to help explain certain concepts. Muller's approach in this book is to provide a lot of details, but the book remains very readable, as long as the reader focuses on the information that they want to glean. I suspect that most readers will get the general idea of a section and then skim over many of the details, unless that specific energy source is something they deeply wanted to know about.

For example, Chapter 16: Electric Automobiles focuses on the possibility of having automobiles that are run entirely on electricity. This is something that Muller clearly doesn't put much faith in, as the section title "The Electric Auto Fad" makes clear. And once you've grasped his view on the matter, there really isn't much reason for the reader to read the one-page detailed descriptions of the Tesla Roadster, Chevy Volt, or Nissan Leaf . . . unless you absolutely want to know more about those particular vehicles, or about electric autos in general. (I felt perfectly comfortable skimming those pages.)

Richard Muller was known by many as one of the most prominent "global warming skeptics". His work with the Berkeley Earth Surface Temperature Project③ was funded largely by sources that are identified as conservative, so it was viewed with a great deal of suspicion within the environmental activism community. The results came back in support of the scientific consensus, as Muller describes in the book: "we concluded that none of the legitimate concerns of the skeptics had improperly biased the prior results". Those who were concerned about global warming were quite excited. (See the response from our very own About.com Green Living Guide for an example.)

Though Muller acknowledges the reality of global warming and humanity's role in it, this doesn't mean that Muller is fully on board with the rhetoric of the environmental left. As he says in the book, "none of the well-known proposals to stop global warming that have been made have any realistic chance of working, even if they are fully implemented".

As a result, this presents a very different book on energy policy from many others. Not only is it written by a scientist, but it's written by a scientist who views the global warming skeptical movement to be rooted in very legitimate concerns... concerns that he feels he's helped lay to rest. Anyone who's interested in energy policy will likely find Muller's perspective a unique one that differs from the other sources they've gotten from, which are often enmeshed firmly in only one side of the debate.

Notes:

① Richard A. Muller 理查德·A. 穆勒(1944年1月6日—),美国加州大学伯克利分校物理学教授

② shale oil 页岩油

③ Berkeley Earth Surface Temperature Project 伯克利地表温度项目

Exercises

Work in groups and discuss the following questions.

1. What is the section "The Energy Landscape" about?
2. What is the title for Muller's book?
3. How many sections in this book?
4. What is the topic of Chapter 16?
5. Is Richard A. Muller a scientist?

Unit 11 Petrochemistry

Text A Crude Oil

Crude oils[①] are compounds that mainly consist of many different hydrocarbon compounds that vary in appearance and composition. Average crude oil composition is 84% carbon, 14% hydrogen, 1%–3% sulphur, and less than 1% each of nitrogen, oxygen, metals and salts.

Crude oils are distinguished as *sweet* or *sour*, depending upon the sulphur content present. Crude oils with a high sulphur content, which may be in the form hydrogen sulphides, are called *sour*, and those with less sulphur are called *sweet*.

A process called fractional distillation separates crude oil into various segments. Fractions at the top have lower boiling points than fractions at the bottom. The bottom fractions are heavy, and are thus "cracked" into lighter and more useful products.

Directly from the well, raw or unprocessed ("crude") oil is not useful. Though light sweet oil has been used directly as a burner fuel, these lighter fragments form explosive vapors in fuel tanks, and thus are dangerous. The oil must be separated into various parts and refined before used in fuels and lubricants, and before some of the by-products form materials such as plastics, detergents, solvents, elastomers, and fibers such as nylon and polyesters.

Crude oil and natural gas are extracted from the ground, on land or under the oceans, with oil wells. Ships, trains, and pipelines transport extracted oils and gasses to refineries[②].

Refineries then execute processes that cause various physical and chemical changes in the crude oil and natural gas. This involves extremely specialized manufacturing processes. One of the important processes is distillation, i. e. separation of heavy crude oil into lighter groups (called fractions) of hydrocarbons. There are two processes of distillation: CDU process and VDU process. The objective of the CDU process is to distill and separate valuable distillates (naphtha, kerosene, and diesel) and atmospheric gas oil (AGO) ③ from the crude feedstock. The technique used to carry out the above process is called complex distillation. On the other hand, the objective of the VDU process is to recover valuable gas oils from reduced crude via vacuum distillation. Two of the fractions of distillation are fuel oil and naphtha, which are familiar to consumers. Fuel oil is used for heating for diesel fuel in automotive applications. Naphtha is used in gasoline and also used as the primary source for petrochemicals.

Refining is the processing of one complex mixture of hydrocarbons into a number of other complex mixtures of hydrocarbons. Refining is where the job of oil industry stops and that of petrochemical industry starts. The raw materials used in the petrochemistry industry④ are known as feedstocks. These are obtained from the refinery: naphtha, components of natural gas such as butane, and some of the by-products of oil refining processes, such as ethane and propane. These feedstocks then undergo processing through an operation called cracking. Cracking is defined as the process of breaking down heavy oil molecules into lighter, more valuable fractions. There are two kinds: steam cracking and catalytic cracking. In steam cracking, high temperatures are used. Catalytic cracking is when a catalyst is being used. The plant where these operations are conducted is called a "cracker". Once these operations complete, new products are obtained that serve as building blocks of the petrochemical industry: olefins, i. e. mainly ethylene, propylene, and the so-called C4 derivatives⑤, including butadiene—and aromatics, so called because of their distinctive perfumed smell, i. e. mainly benzene, toluene and the xylenes.

Then petrochemicals go through various processes that eventually contribute to the final output of products like plastics, soaps and detergents, healthcare products like aspirin, synthetic fibres for clothes and furniture, rubbers, paints, insulating materials, etc.

The global demand for petrochemical products continuously rises. One of the major concerning issues in today's world is the dependence of the modern society on oil and gas and various other petroleum products. Besides this, there are problems relating to the increasing scarcity of workable hydrocarbon deposits and global warming. Thus, solutions must be found in the next few years, such as making more efficient use of the energy available and use more renewable energy sources in addition to hydrocarbons.

Notes:

① crude oil 原油;毛油

② refinery 炼油厂;精炼厂

③ atmospheric gas oil (AGO) 常压瓦斯油;常压柴油

④ petrochemistry industry 石油化学工业

⑤ C4 derivative C4 衍生品

Exercises

Ⅰ. Read each of the following statements carefully and decide whether it is true or false according to the text.

1. Average crude oil composition is 80% carbon, 18% hydrogen, 1%–3% sulphur, and less than 1% each of nitrogen, oxygen, metals and salts. ( )
2. Crude oils are distinguished as sweet or sour, depending upon the sulphur content present. ( )
3. A process called fractional distillation separates crude oil into various segments. ( )
4. Fractions at the top have higher boiling points than fractions at the bottom. ( )
5. Naphtha is used in gasoline and also used as the primary source for petrochemicals. ( )

Ⅱ. Answer the following questions according to the text.

1. What is crude oil?
2. What is the average crude oil composition?
3. What is the standard to distinguish crude oils?
4. What is refining?
5. According to the text, what is one of the major concerning issues in today's world?

Ⅲ. Translate the following terms into their Chinese or English equivalents.

1. crude oil
2. hydrocarbon compounds
3. sulphur content
4. hydrogen sulphides
5. fractional distillation
6. 爆炸性气体
7. 副产品
8. 制造工艺
9. 原油原料
10. 复杂精馏

Ⅳ. Translate the paragraph into Chinese.

Directly from the well, raw or unprocessed ("crude") oil is not useful. Though light sweet oil has been used directly as a burner fuel, these lighter fragments form explosive vapors in fuel tanks, and thus are dangerous. The oil must be separated into various parts and refined before used in fuels and lubricants, and before some of the by-products form materials such as plastics, detergents, solvents, elastomers, and fibers such as nylon and polyesters.

Text B Petrochemistry

It may be possible to make petroleum from any kind of organic matter under suitable conditions. The concentration of organic matter is not very high in the original deposits, but petroleum and natural gas evolved in places that favored retention, such as sealed-off porous sandstones. Petroleum, produced over millions of years by natural changes in organic materials, accumulates beneath the earth's surface in extremely large quantities.

The first oil commercial was set up in 1859, two years after which the first oil refinery was set up. The industry grew in the late 1940s. Demand for products from the petrochemical industry grew during the World War II. The demand for synthetic materials① increased, and this rising demand was met by replacing costly and sometimes less efficient products with these synthetic materials. This caused petrochemical processing② to develop into a major industry.

Before this, petrochemical industry was a tentative sector where various experiments could be carried out. The industry used basic materials: synthetic rubbers in the 1900s, Bakelite, the first petrochemical-derived plastic in 1907, the first petrochemical solvents③ in the 1920s, polystyrene in the 1930s. After that period, the industry produced materials for a large variety of areas—from household goods (kitchen appliances, textile, furniture) to medicine (heart pacemakers, transfusion bags), from leisure (running shoes, computers) to highly specialized fields like archaeology and crime detection.

Petrochemistry is a science that can readily be applied to fundamental human needs, such as health, hygiene, housing and food. To many, this comes as a surprise. The word "chemistry" itself conjures up a world of mystery—what it really does is very much taken for granted. Yet it is a fascinating science and an inventive business sector, constantly adapting to new environments and meeting new challenges.

Chemicals derived from **petroleum** or **natural gas—petrochemicals**—are an essential part of the chemical industry today. However, all this is little known. **Petrochemicals do not reach the final consumer**—the man in the street; they are first sold to customer industries, undergo several transformations, and then go into products that seem to bear no relation whatsoever to the initial raw material. As a result, few of us make the connection between the **petrochemical industry** and their GP's equipment, their DVDs, food packaging or computers; few realize the amount of scientific efforts that went into these commonplace objects. Although benefiting daily from end products that have been made thanks to the input of the petrochemical industry, more often than not we see no obvious connection between these everyday commodities and petrochemistry.

Notes:

① synthetic material 合成材料人造材料合成物

② petrochemical processing 石油化学加工

③ petrochemical solvent 石油化工溶剂

Exercises

Work in groups and discuss the following questions.

1. Can nature produce petroleum?
2. What is the reason for petrochemical processing to develop into a major industry?
3. When did the first oil commercial be set up?
4. What are the basic materials used in the petrochemical industry?
5. Can petrochemistry readily be applied to fundamental human needs?

Unit 12 Aeronautics

Text A NASA's Chief Scientist: The Future of Space Exploration Is International Partnerships

Ellen Stofan, NASA's chief scientist, saw her first rocket launch at age 4, thanks to her father's job at NASA as an engineer. But at a Future Tense film screening of *The Dish* in Washington, D. C. , last week, Stofan said that for many people she meets, what first sparked a space obsession was the Apollo program—President John F. Kennedy's audacious commitment in 1961 to putting Americans on the moon before the end of the decade.

Today, NASA's goal to put astronauts on Mars by the 2030s could be a similarly unifying project. And not only in the United States, exploration in the 21st century is likely to be a far more globally collaborative project than it was during the fierce Cold War Space Race between the United States and Soviet Union.

Why has the idea of reaching Mars captured the world? A trip to Mars is a priority for many scientific reasons—some believe it's the planet that most resembles our own, and one that could answer the age-old question of whether we're alone in the universe. But, as Stofan noted, there's also been a long popular fascination with the planet. Ever since Giovanni Virginio Schiaparelli first observed the *canali* on Mars in the 1800s or when H. G. Wells wrote about aliens from Mars in his 1898 science fiction novel, *The War of the Worlds*①, the planet has loomed large in the public's imagination.

And perhaps it's this historic obsession that partly explains the more international effort: The US is hardly the only country dreaming of deep space—and a trip to Mars—these days. India has plans to put astronauts in the sky, Japan just launched a spacecraft to collect asteroid samples, and of course, the European Space Agency had the recent Rosetta mission and Philae lander. It seems that what Apollo did for America's imagination and spirit of invention, foreign space programs can also do domestically. "You see countries like India really investing in their space program because they see it as inspirational and good for their economy," Stofan told the audience.

The truth is, as Stofan put it, "When we go to explore, we do it as a globe." In a conversation outside the event, she recounted the stories of some of the astronauts featured in the 2007 documentary, *In the Shadow of the Moon*②, who travelled the world after they returned from the Apollo missions of the 1960s and '70s. People from all sorts of countries welcomed them, not just as Americans, but as "our astronauts." "People see space as a place where you go and cooperate," she told me.

This spirit of trans-border ownership and investment seems set to continue. One key part of this is the Global Exploration Roadmap, an effort between space agencies like NASA, France's Centre National d'Etudes Spatiales, the Canadian Space Agency, and the Japan Aerospace Exploration Agency, among many others. The partnership is intended to aid joint projects from the International Space Station to expeditions to the Moon and near-Earth asteroids— and of course, to reach Mars. On a recent trip to India's space agency, Stofan recounted to me, she met with many Indian engineers who were just as excited as the Americans to get scientists up there, not only to explore, but also to begin nailing down the question of whether there was ever life on the red planet.

It's also clear that the next stage of space exploration will not only be more global, but will equally involve greater private and public partnerships. Companies like Space X and Boeing are increasingly involved in NASA's day-to-day operations, including a joint project that could carry astronauts into space in 2017. NASA's view is to turn over to the private sector those projects that in a sense have become routine, Stofan suggested, and let NASA focus its resources on getting to Mars.

This environment feels a lot different from the secretive and adversarial Space Race days, when the US and Soviet Union battled to reach the moon first. The Cold War is over, of course, but with it, the funding commitment may also be missing this time around. Stofan mentioned, in response to an audience question, that at the time of the Apollo missions. NASA received up to about 4 percent of the federal budget, while now it's only around 0.4 percent. Between its peak in 1966 and 2014, the "space flight, research, and supporting activities" section of the budget has contracted significantly, according to Adam Rosenberg, a policy analyst with the Committee for a Responsible Federal Budget.

The dollars are still large, of course, but perhaps increased international and private cooperation can be seen as an efficient, clever way to do more with less.

So, what does the future hold? NASA is extremely focused on how to get to Mars and back again safely, Stofan told the audience. But the fun rule of science fiction, she suggested, is to start envisioning what the steps after that might be. For example, what it might be like to live on Mars? After all, science often gets its inspiration from the creative world. Just look at how similar mobile phones are to the communicators from *Star Trek*, she pointed out, or the fact that MIT students made real-life version of the robotic sphere that Luke Skywalker trains with in Star Wars. "Stories are a great counterpoint to science."

What would Stofan like to see on the big screen next? "*The Martian*, I think it's being made into a movie in already. And I wish someone would redo *Dune*."

Notes:

① *The War of the Worlds*《星际战争》,又名《世界大战》

② *In the Shadow of the Moon* 本文中指纪录片《月之阴影》

Exercises

Ⅰ. Read each of the following statements carefully and decide whether it is true or false according to the text.

1. It was the Apollo program that sparked a space obsession. (　)
2. H. G. Wells wrote about aliens from Mars in his 1998 science fiction novel. (　)
3. All sorts of countries intend to explore the Mars and make clear whether there was ever life on it. (　)
4. The next stage of space exploration will be more global. (　)
5. Increased international and private cooperation can be seen as an efficient, clever way to conduct space exploration with fewer budgets. (　)
6. Many countries believe that space program can be a stimulus to a country's economy. (　)

Ⅱ. Answer the following questions according to the text.

1. Why so many countries are obsessed with the idea of reaching Mars?
2. Why the next stage of space exploration will involve greater private partnerships?
3. What will NASA's future space exploration be focused on?
4. What's the difference between the former Space Race days and the 21st century space exploration?
5. What countries were involved in the Cold War Space Race?

Ⅲ. Translate the following terms into their Chinese or English equivalents.

1. space exploration
2. Apollo program
3. space race
4. federal budget
5. Martian
6. 欧洲空间局
7. 小行星
8. 太空飞行
9. 火星
10. 科幻小说

Ⅳ. Translate the paragraph into Chinese.

It's also clear that the next stage of space exploration will not only be more global, but will equally involve greater private and public partnerships. Companies like Space X and Boeing are increasingly involved in NASA's day-to-day operations, including a joint project that could carry astronauts into space in 2017. NASA's view is to turn over to the private sector those projects that in a sense have become routine, Stofan suggested, and let NASA focus its resources on getting to Mars.

Text B A "Starshade" Could Help NASA Find Other Earths Decades Ahead of Schedule

A next-generation space telescope is in the works—but if it is to see potentially habitable planets, it will need to block out their suns.

Credit: COURTESY OF NASA/JPL-CALTECH

Can a next-generation NASA① space telescope take pictures of other Earth-like planets? Astronomers have long dreamed of such pictures, which would allow them to study worlds beyond our solar system for signs of habitability and life. But for as long as astronomers have dreamed, the technology to make it happen has seemed many decades away. Now, however, a growing number of experts think NASA's Wide-Field Infrared Survey Telescope (WFIRST)② could take snapshots of "other Earths"—and soon. The agency formally started work on the observatory in February of this year and plans to launch it in 2025.

When WFIRST launches, it will sport a 2.4-meter mirror that promises panoramic views of the heavens and will use its wide eye to study dark energy, the mysterious force driving the universe's accelerating expansion. But another hot topic—the existential quest to know whether we are alone in the universe—is already influencing the mission.

Researchers have discovered more than 3,000 planets around other stars and expect to find tens of thousands more within the next decade. Rough statistics suggest that every star in the sky is accompanied by at least one such exoplanet and that perhaps one in five sunlike stars bears a rocky orb in a not-too-hot, not-too-cold "habitable zone" where liquid water can exist. The best way to learn whether any of these worlds are Earth-like is to see them, but taking a planet's picture from light years away is far from easy. A habitable world would be a faint dot lost in the overpowering glare of its larger, 10-billion-time-brighter star.

Earth's turbulent, starlight-blurring atmosphere is also a severe obstacle to imaging faint planets from ground-based observatories, and most experts agree that the solution is to use space telescopes. But neither NASA's Hubble Space Telescope nor its supersize successor, the James Webb Space Telescope set for launch in 2018, comes close to achieving the high contrast needed. To help capture planetary shots, WFIRST will have an advanced planet-imaging coronagraph③, an instrument inside the telescope that filters out starlight using a complex series of masks, mirrors and lenses. But this instrument was a late addition to WFIRST, which is not optimized for a coronagraph. Consequently, most experts predict that its coronagraph will fall short of the contrast required to image other Earths. Indeed, snapping such images is so challenging that NASA's tentative plans call for putting it off for perhaps 20 years or more as the agency develops the technology and budgetary breathing room to build an entirely new space telescope after WFIRST.

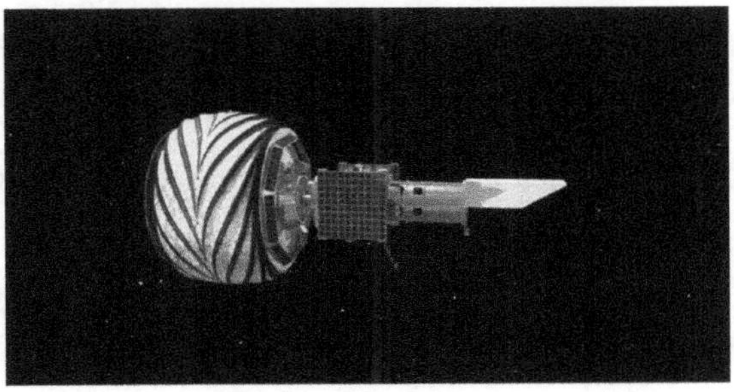

These sequential images show how a starshade could be deployed with a future space telescope. Initially folded up for launch into space (1), the starshade would detach and unfurl (2) and then fly away (3) to its station tens of thousands of kilometers ahead of the telescope.

A device called a starshade④ might offer a shortcut. A starshade is a sunflower-shaped, paper-thin screen half as big as a football field that would float tens of thousands of kilometers directly ahead of WFIRST, blocking out a target star's light in much the same way one might blot out the sun in the sky with an extended thumb. Because starshades work with practically any telescope, one on WFIRST could cast a deeper shadow and see fainter planets than a coronagraph. Working in tandem, the starshade and the telescope could take pictures of perhaps 40 planets, including a few that in size and orbit would mirror Earth. "If and only if it had a starshade, WFIRST could give us images of a few true-blue Earths late next decade rather than waiting for another 20 years," says Jeremy Kasdin, a Princeton University professor and lead scientist for WFIRST's coronagraph. "This is a real opportunity to find another Earth sooner and for less money before making a huge investment in NASA's next giant space telescope."

Despite WFIRST being nearly a decade away from launch, the decision to move forward with preparations for a starshade rendezvous must come soon because WFIRST must receive minor modifications to allow it to sync up with a starshade across tens of thousands of kilometers of empty space. As such, an official starshade mission does not exist. Instead Paul Hertz, director of NASA's astrophysics division, says the agency is "in a 'don't preclude a starshade' mode." So far not precluding a starshade closely resembles a concerted effort to build one: when NASA first announced the formal start of WFIRST, it also confirmed that the telescope would be launched into an orbit 1.5 million kilometers from Earth, where conditions are tranquil enough for a starshade to function. In addition, the agency recently formed the StarShade Readiness Working Group and officially designated the starshade as a "technology development activity"—moves that could accelerate the agency's progress.

In fact, in the basement of Princeton's sprawling Frick Chemistry Laboratory, Kasdin is already working on a test bed: a meter-wide, 75-meter-long tube with a camera at one end, a laser at the other and a scaled down starshade in between. By the end of the summer, he predicts, the test bed will have demonstrated the necessary contrast ratio that, scaled up to full size, could enable the imaging of Earth-like planets. Meanwhile aerospace company Northrop Grumman has tested miniaturized starshades at a dry lakebed in Nevada and at a giant solar telescope in Arizona. And at the NASA Jet Propulsion Laboratory, researchers are demonstrating how to fabricate a larger-scale starshade's delicate petals, fold the entire structure up inside a rocket, and deploy and unfurl it to the size of a baseball diamond⑤.

Not all the obstacles to a starshade are technological. One for WFIRST could easily cost a billion dollars—far too much extra money for the telescope's budget to bear. Consequently, it would have to first be proposed and approved as an independent project with its own substantial supply of NASA funding. That's a high hurdle for a still nascent technology to clear, but the payoff could be historic: delivering the first image of an alien Earth is an event that can happen only once. Should we try to do it as quickly as possible or delay it for decades more? NASA and the astronomical community must decide soon.

Notes:

① NASA National Aeronautics and Space Administration 美国国家航空航天局

② Wide-Field Infrared Survey Telescope (WFIRST) 广域红外勘测望远镜

③ coronagraph 日冕仪

④ starshade 遮星板

⑤ a baseball diamond 棒球场

Exercises

Work in groups and discuss the following questions.

1. What can astronomers do with NASA's Wide-Field Infrared Survey Telescope (WFIRST)?
2. What are the obstacles to taking pictures of other Earth-like planets?
3. What is a starshade and how does it work?
4. How many planets have researchers discovered around other stars?
5. How is Kasdin working in the laboratory?

Unit 13 Auto Industry

Text A The Major Problem with Cheap Electric Cars

Mitsubishi is the latest in a long line of automakers to slash prices on an electric car, the unpronounceable, unfortunately named i-MiEV. The model was the cheapest electric vehicle (EV) on the market in 2014, yet it's still hard to imagine many drivers excitedly running out to buy one.

Overall, American consumers are paying more for cars. The average price paid for a new light vehicle in November 2014 was $ 32,769, up 3% compared with October, and 1.1% year over year, according to Kelley Blue Book①. Yet as average prices rise, one segment of the new-car market has gotten markedly cheaper. "Electric vehicles had the largest decrease in pricing, down 15% due to several price cuts during the past year," says Karl Brauer, senior analyst for Kelley Blue Book.

This is a trend that only seems to be growing, as another electric car, a golf-cart-like-model Mitsubishi, has just been marked down by 20%.

In the world of electric vehicles, Mitsubishi's i-MiEV has mostly been an afterthought. The model lacks the sex appeal of Tesla, it's not nearly as practical as the plug-in Chevy Volt (which has a driving range of 380 miles, or 612 km, thanks to battery and gas power), and the large price cut of the Nissan Leaf② meant that it offered far more bang for the buck③ than Mitsubishi's electric car. The Leaf also generally has received far better reviews, and it has superior range: 75 miles (121 km) on a single charge, compared with about 60 miles (97 km) for the i-MiEV.

The best indication of interest in specific electric-car models is simply the sales tally. Tesla sold 21,500 Model S cars by the end of 2013. Data collected by Autoblog④ indicates that

Nissan and Chevrolet were neck and neck for the EV sales lead that year, selling a bit more than 20,000 Leafs and Volts through November. And what about Mitsubishi's EV? *Automotive News* reported that a grand total of 12 i-MiEVs sold in the month. Just over 1,000 i-MiEVs were purchased nationally in 2013, and many of them were sold with the help of dealer incentives amounting to discounts of up to \$ 10,000.

In light of data like that, and the examples set by Nissan, Chevy and other automakers that saw sales take off after electric model price cuts, it's understandable that Mitsubishi just announced a major price slashing for the next i-MiEV. The 2014 model starts with a sticker price of \$ 23,845, a drop of \$ 6,130 from the previous edition. Federal tax credits effectively knock the price paid out of pocket by drivers down to \$ 16,345, and in states with further incentives like California, the net takeaway price potentially inches down to under \$ 14,000. That's among the cheapest prices available for any new car sold, no matter how it's powered.

Hopefully, the price cut will help Mitsubishi sell more than a dozen EVs per month. But even at that price, it's not clear if all that many drivers will bite. The reviewers at *Consumer Reports* say the i-MiEV is "not a car in which anyone will be happy spending time," and that the newly discounted price is "still a lot of money for a car that feels like little more than an enclosed golf cart. The appeal lies solely in providing attainable access into the world of pure-electric cars. At this price, it becomes more feasible as a second, occasional-use car."

Notes:

① Kelley Blue Book 一家汽车评价公司，成立于1918年，总部位于美国加利福尼亚州欧文市。2013年，该机构与易车网合作，进入中国

② Nissan Leaf 一款5座掀背两厢的纯电动汽车，全方位实现零排放移动模式与环保创新。该车配备了先进的锂离子电池驱动的车辆底盘，行程达160公里，可以充分满足消费者在实际生活中的驾驶需求

③ bang for the buck 划算，货真价实

④ Autoblog 一个提供汽车和汽车行业资讯的新闻网站，其涵盖内容广泛，包括新闻、评论、播客、清晰的图片摄影和评注等

Exercises

Ⅰ. Read each of the following statements carefully and decide whether it is true or false according to the text.

1. I-MiEV was the cheapest and most popular EV on the market in 2014. (　　)

2. EVs take up one segment of the new-car market. ( )
3. Even though i-MiEV lacks the sex appeal of Tesla, it is as practical as the plug-in Chevy Volt. ( )
4. Nissan Leaf enjoys more advantages than Mitsubishi's electric car. ( )
5. The sales tally can best indicate buyers' interest in specific electric-car models. ( )
6. I-MiEV not only provides attainable access into the world of pure-electric cars. ( )

Ⅱ. Answer the following questions according to the text.

1. Why doesn't Mitsubishi's i-MiEV enjoy advantages in the world of electric vehicles?
2. Why are electric vehicles becoming markedly cheaper?
3. How much is electric vehicles' decrease in pricing according to Kelly Blue Book?
4. What are the best-sellers in the EV market in 2013?
5. What is the major problem with cheap electric cars?

Ⅲ. Translate the following terms into their Chinese or English equivalents.

1. electric vehicle
2. sticker price
3. sales tally
4. enclosed golf cart
5. occasional-use car
6. 轻型车辆
7. 插电型车辆
8. 裸车售价
9. 纯电动汽车
10. 联邦税收抵免

Ⅳ. Translate the paragraph into Chinese.

Hopefully, the price cut will help Mitsubishi sell more than a dozen EVs per month. But even at that price, it's not clear if all that many drivers will bite. The reviewers at *Consumer Reports* say the i-MiEV is "not a car in which anyone will be happy spending time," and that the newly discounted price is "still a lot of money for a car that feels like little more than an enclosed golf cart. The appeal lies solely in providing attainable access into the world of pure-electric cars. At this price, it becomes more feasible as a second, occasional-use car."

Text B The Dangers of an Exploding Car Battery

Automotive electrical systems aren't tremendously complicated, when you look at the big picture, and a lot of the technologies we use today—from alternators① to lead acid batteries②—have been around for a long time, but there are still a lot of folks out there who look askance at a relatively simple task like hooking up jumper cables③, possibly because they've heard that doing it wrong can cause a lot of damage, or even cause the battery to blow up. And while you'll find that a lot of weird myths and rumors about automotive technology are just that—unsubstantiated myths and rumors—the dangers associated with hooking up jumper cables, or a battery charger, incorrectly can cause a lot of damage, or even result in an exploding battery. The good news is that if you take the time to understand why a car battery can explode, and take a few basic precautions, this isn't a problem you'll ever have to worry about.

Safely Connecting Jumper Cables or a Battery Charger

There are a few general rules of thumb that can help you safely connect jumper cables, but there are also a number of special cases that supersede those rules.

So before you use your car to provide a jumpstart, accept a jump from someone else, or hook up a charger to your battery, the first thing you need to do is check your owner's manual to make sure your car doesn't have designation connection points other than your battery. If your car has a battery that's buried somewhere weird, like a wheel well or the trunk, then there's a good chance that you're supposed to use a junction block or some other kind of remote connection.

Regardless of the vehicles in question, the basic idea behind safely connecting jumper cables is to connect the electrical system of a donor vehicle, which has a good battery, to the electrical system of a vehicle with a dead battery. Positive should be connected to positive, and negative should be connected to negative, as connecting backwards can damage both vehicles and create potentially hazardous sparks, but more on that later.

In most cases, the safest procedure for safely hooking up jumper cables is to:

1. Ensure that the keys of both vehicles are in the "off" position.

2. Connect one jumper cable to the positive (+) terminal of the donor battery.
3. Connect the same cable to the positive (+) terminal of the dead battery.
4. Connect the other jumper cable to the negative (–) terminal of the donor battery.
5. Connect the other end of that cable to bare metal on the engine or frame of the vehicle with the dead battery.

Connecting a battery charger is done in much the same way, except instead of a donor battery, you're using a charger. The positive charger cable should be connected to battery positive (+), after which the negative charger cable should be connected to bare metal on the engine or frame of the vehicle.

There are some exceptions where positive is ground, but in most automotive electrical systems, negative is ground. That's why you can connect a charger, or a jumper cable, to bare metal on the frame or engine of a vehicle with a dead battery and have current flow into the battery. Of course, it's technically possible to connect directly to battery negative, and it may even be easier in some cases. So if it's possible, and it's essentially the same thing as connecting to some other ground, why go through the trouble? Because you don't want your battery to explode.

The Science of Exploding Car Batteries

Car batteries are referred to as "lead acid" because they utilize plates of lead submerged in sulfuric acid to store and release electrical energy. This technology has actually been around since the 18th century, and it isn't terribly efficient from either an energy-to-weight or energy-to-volume standpoint. However, they do have an excellent power-to-weight ratio, which essentially just means that they are good at providing the high levels of on-demand current required by automotive starters.

The downside of lead acid batteries, other than the fact that they aren't a terribly efficient way to store energy, is that they're made up of fairly hazardous materials, and those hazardous materials can interact in dangerous ways. The presence of lead is the primary reason that car batteries have to be carefully and properly disposed of, and the presence of sulfuric acid is why you have to take care when handling them, unless you want holes eaten in your clothes or chemical burns on your skin.

Of course, the danger that we're particularly concerned with here is a sudden and catastrophic explosion, and the source of that particular hazard flows from the interaction between the lead and sulfuric acid in a battery. Small amounts of hydrogen gas are produced during both the process of discharge and during charging, and hydrogen is highly flammable. So when a battery has discharged to the point where it can no longer power a starter motor, there's always a chance that some amount of hydrogen gas is still lingering inside the battery, or leaking out of the battery, just waiting for an ignition source. The same is true of a battery that has just been charged, as charging—and particularly overcharging—with a high voltage leads to the formation of both oxygen and hydrogen.

Preventing Car Battery Explosions

There are two primary ignition sources that you have to worry about, and they can both be avoided with careful charging, jumping, and maintenance practices. The first ignition source is a spark created when connecting or disconnecting a jumper or charging cable. This is why it's so important to connect to bare metal on the engine or frame instead of the battery. If you hook up the negative jumper cable to the battery itself, any lingering hydrogen may be ignited by the ensuing spark. This is also why it's a good idea to wait to turn on, or plug in, your charger until after it's connected.

The other type of car battery explosion still involves hydrogen gas, but the ignition source is inside the battery. The issue is that if a battery isn't properly maintained, and the electrolyte level is allowed to drop, the lead plates will be exposed to oxygen and may warp. This can lead to the plates flexing and touching during the extreme current drain initiated whenever you crank the starter motor, which can result in a spark inside the battery. That, in turn, can ignite any hydrogen present in the cell, causing the battery to explode.

What About Sealed Car Batteries?

There are two main types of sealed car batteries: traditional lead acid batteries that just aren't serviceable, and VRLA (valve-regulated lead acid) batteries that actually don't need to be serviced. In the case of VRLA batteries, the electrolyte is contained in a saturated glass mat or gelled, so evaporation isn't really an issue, and there really is no need to ever add more electrolyte, and there is little or no danger of the plates ever becoming exposed to the air. Sealed batteries that use liquid electrolyte, however, can cause issues later in life.

If you have a VRLA battery, be it absorbed glass mat or gel cell, then the chances of the battery ever exploding are exceedingly low. However, it's still a good idea to follow jumpstart and charging best practices just so that you don't get out of the habit. However, maintenance of these batteries is actually impossible, so you don't have to worry about checking the charge or electrolyte level[4] regularly.

Special care should be taken with non-VRLA sealed and "maintenance-free" batteries, since at least some level of evaporation will take place over time, and the situation will only get worse if the battery is allowed to fully discharge repeatedly, or overcharged with a high voltage. So while it's a good idea to be careful around any battery when jump-starting or charging it, it's an even better idea to be extra careful when dealing with old, discharged, or recently charged non-VRLA sealed batteries.

Notes:

① alternator 交流发电机

② lead acid battery 铅酸电池

③ jumper cable 跨接电缆

④ electrolyte level 电解液液位

Exercises

Work in groups and discuss the following questions.

1. What's the basic idea behind safely connecting jumper cables?
2. Why are car batteries referred to as "lead acid"?
3. What's the disadvantage of lead acid batteries?
4. What are the primary ignition sources?
5. How many types of sealed car batteries are there? What are they?

Unit 14 Communication and Transportation

Text A Apple Watch Retrospective: One Year Later

It's been a little over a year since Apple officially released the Apple Watch, its first foray into the wearable space. Now as we anticipate a new model (which will likely be announced next month during Apple's annual Worldwide Developers Conference, WWDC①), we thought it might be fun to take a look back at the past year, how we've used the Apple Watch, and how the device has changed.

The Beginning

The Apple Watch launched with a great deal of fanfare, with just three versions: The Apple Watch, Apple Watch Sport, and Apple Watch Edition. Announced several months before the device was actually available, preorders far exceeded the initial supply of the device. That meant that many buyers, even those who placed their order the day the watch was announced, had to wait several weeks from the initial launch day (if not longer) to actually get the Apple Watch on their wrist.

Those who didn't pre-order device travel to the Apple Store for try-on appointments, special 15-minute appointments in the store where you could try out different versions of the Apple Watch, and get a walkthrough of some of its features.

Perhaps the most intrigue came from the Apple Watch Edition, the high-end version of the Apple Watch which has a $ 10,000 price tag. From what we can tell, Apple didn't end up selling a ton of that version of the watch (nor did it expect to). That said, the coveted $ 10,000 Watch did get a lot of interest, with some Apple Watch buyers even trying to figure out how to make their own Apple Watch gold, going the impression that their $ 350–$ 500 buy was actually the more expensive model.

Software Updates

Despite the fact we're still using the same Apple Watch model that was initially released, there have been quite a few updates to the device over the past year that have improved its software, and made it an even better device to use.

We're currently using watchOS 2. 2, the second full version of the Apple Watch's software which was released in March during an event where the company also announced a new version of the iPad Pro as well as a new, less expensive version of the iPhone called the iPhone SE.

Since its initial launch, the Apple Watch software gained a number of significant improvements from software updates. The largest improvement perhaps happened in when Apple started to allow the support of native apps on the Watch. While most of the Apple Watch's apps still are run from your iPhone, native app support means that some of your Apple Watch's apps are able to function even if you decide to leave your smartphone behind.

Other improvements include new watch faces including a 24-hour timelapse② options, a nightstand mode③, which allows your Apple Watch to be readable while you're charging it at night, the ability to draw in multiple colors, and improvements to the Maps and Apple Pay app. While we're still using the same hardware a year later, those software improvements in a way have made the Apple Watch seem like a slightly different device, giving it a bit of an improved feel while keeping the same packaging.

New Accessories

The Apple Watch launched with a limited number of watch bands ranging from the Apple Watch Sports band to leather and a Milanese loop option, with your first watch band purchase made in conjunction with your purchase of the watch itself. Since then, a number of third-party manufacturers have released their own bands for the Apple Watch include a few designers. Apple has also started to release seasonal lineups of its traditional bands, offering new color choices along the way. Most recently during an event in March, it unveiled a new line of nylon bands.

The nylon band represents a new look for Apple's Apple Watch bands, and has a "four layer construction." It will be available in seven different colors: gold/red woven nylon, gold/royal blue woven nylon, royal blue woven nylon, pink woven nylon, pearl woven nylon, scuba blue woven nylon, and black woven nylon.

More than just watch bands, this year we've seen a number of other interesting accessories released for the Apple Watch, particularly when it comes to charging the device. At launch, Apple only offered a charging cable. A number of third-party manufacturers came up with interesting charging docks④ and cases for the watch, eventually prompting Apple to release a charging dock of its own for the Apple Watch as well.

Designer Partnerships

Over the past year Apple also announced a partnership with Hermès, a move that unveil a new, slightly different version of the Apple Watch along with the fashion brand. The Hermès versions of the Apple Watch come sporting an exclusive Hermès watch face, and exclusive watch bands unique to the collection.

"Uncompromising craftsmanship. Pioneering innovation. Groundbreaking functionality. Apple Watch Hermès is the culmination of a partnership based on parallel thinking, singular vision, and mutual regard. It is a unique timepiece designed with both utility and beauty in mind. With leather straps handmade by Hermès artisans in France and an Hermès watch face reinterpreted by Apple designers in California, Apple Watch Hermès is a product of elegant, artful simplicity—the ultimate tool for modern life," read the page on Apple's website announcing the collaboration.

The watch comes in a "Double Tour" model with an extra long band designed to wrap around the wrist several times, as well as a single tour version, which looks closer to a traditional watch band. The Double Tour version starts at $ 1,250, while the Single Tour version starts at $ 100. Both uses the same body as the traditional Apple Watch, so besides the unique watch faces, you're shelling out that extra cash specifically for the watch bands.

Beyond partnerships with Apple itself, designer Rebecca Minkoff decided to partner up with Case-Mate to launch a new line of bands for the Apple Watch as well. Her bands include a double-wrap leather band priced at $ 100, and a debossed chevron band, which is priced at $ 80.

Amazing Apps

Quite a few developers have created some pretty amazing stuff for the Apple Watch, which have really helped the platform shine over the past year. From interesting games (and old favorites like Pong), to apps that can actually make a big difference in how we live.

Some of the most interesting apps that have been released for the Apple Watch over the past have come in the form of healthcare apps. The apps are being used to help some cancer patients with their treatments, by monitoring those patients while they're at home. By being able to track those patients relatively 24/7 as far as their physical and emotional condition, doctors are able to get a full picture on how a treatment is effecting someone.

The Apple Watch is also being used in treatments for epilepsy⑤ patients. The study, which is being conducted by the Johns Hopkins University School of Medicine, has patients take daily surveys and make journal entires about their disease, and tries to get them to document when they have seizures and what happens to their body prior to one come on. Thanks to the Apple Watch's heart rate monitor, accelerometer, and gyroscope, researchers will be able to track changes in heart rate as well as body movement in patients, ultimately gaining a better understanding of the disease.

Hacking

Several ambitious developers have created hacks for the Apple Watch over the past year, transforming the device, and enabling it to do thing that were not possible with the original hardware and software.

Recently one guy managed to get Windows 95 running on the Apple Watch, although it required both a software and hardware hack, and takes an hour to boot up. Earlier this year another developer was able to do something similar, getting a simulator of Apple's OSX Yosemite to run on the wearable.

The Future

We highly expect Apple to unveil the next version of the Apple Watch in the coming months. What that next version looks like; however, is currently anyone's guess. Current rumors point to the newest version of the device sporting a longer-life battery, a thinner design and perhaps a forward-facing camera so it can be used to FaceTime calls.

Rumors this past week also pointed to the next version of the Apple Watch supporting cellular data⑥, a feature that would make it possible for you to use the Apple Watch without having your iPhone with you. Analysts currently expect Apple to formally announce the next version of the device at WWDC in May. A recent survey indicated that the majority of current Apple Watch owners would be interested in an upgrade if and when it becomes available.

According to Fluent, the company behind the survey, three fifths of currently Apple Watch owners are planning on purchasing the newest version when it's revealed to the public, even though at this point no one is quite certain what improvements might be in store with the new model.

Additionally, the study found that just 8 percent of its respondents had purchased the Apple Watch so far (that three fifths number is from that 8% who said they already owned the Watch). The reason? The majority of respondents said the Apple Watch's high price, even after its recent $ 50 price cut, was a deterrent in having them make the purchase.

Notes:

① Worldwide Developers Conference, WWDC 苹果全球开发者大会

② timelapse 延时拍摄

③ nightstand mode 床头钟模式

④ charging dock 充电底座

⑤ epilepsy 癫痫

⑥ cellular data 蜂窝数据，主要为专门的移动数据通信系统所采用

Exercises

Ⅰ. Read each of the following statements carefully and decide whether it is true or false according to the text.

1. Many buyers had to wait several weeks from the initial launch day to actually get the Apple Watch. (   )
2. In March, a new version of the iPad Pro and a new and more expensive version of the iPhone were announced. (   )
3. Apple finally released a charging dock of its own for the Apple Watch. (   )
4. Apple Watch Hermès is the culmination of a partnership based on parallel thinking, singular vision, and mutual regard. (   )
5. The Apple Watch can cure cancer and epilepsy. (   )
6. Most of the Apple Watch owners are going to purchase the newest version when it is revealed to the public. (   )

Ⅱ. Answer the following questions according to the text.

1. What are the different versions of the Apple Watch at the beginning?
2. What are the software updates of the Apple Watch?
3. What are the interesting accessories released for the Apple Watch?
4. What are the Hermès versions of the Apple Watch like?
5. What hacks have been created for the Apple Watch?

Ⅲ. Translate the following terms into their Chinese or English equivalents.

1. price tag
2. software update
3. nightstand mode
4. nylon band
5. heart rate monitor
6. 蜂窝数据
7. 加速计
8. 充电底座
9. 第三方制造商
10. 启动

Ⅳ. Translate the paragraph into Chinese.

Rumors this past week also pointed to the next version of the Apple Watch supporting cellular data⑥, a feature that would make it possible for you to use the Apple Watch without having your iPhone with you. Analysts currently expect Apple to formally announce the next version of the device at WWDC in May. A recent survey indicated that the majority of current Apple Watch owners would be interested in an upgrade if and when it becomes available.

Text B High Speed Trains

High speed trains are a type of passenger train travel that functions at a speed much higher than that of traditional passenger trains. There are different standards of what constitutes high speed trains based on the train's speed and technology used however. In the European Union, high speed trains are that which travels 125 miles per hour (200 km/h) or faster, while in the United States it is those that travel 90 mph (145 km/h) or faster.

History of High Speed Trains

Train travel has been a popular form of passenger and freight transport since the early 20th century. The first high speed trains appeared as early as 1933 in Europe and the US when streamliner trains were used to transport goods and people at speeds of around 80 mph (130 km/h).

In 1939, Italy introduced its ETR 200 train that had routes from Milan to Florence and was capable of traveling at a top speed of 126 mph (203 km/h). Services and further development for ETR 200 stopped with the beginning of World War II.

After WWII, high speed trains again became a priority in many countries. It was especially important in Japan and in 1957, the Romance car 3000 SSE was launched in Tokyo. The Romance car was a narrow gauge train (a narrower area than 4 feet [1.4 m] across between the railroad's rails) and set a world speed record for its ability to travel 90 mph (145 km/h). Shortly thereafter in the mid-1960s, Japan introduced the world's first high volume high speed train that operated with a standard (4 ft) gauge.

It was called the Shinkansen and officially opened in 1964. It provided rail service between Tokyo and Osaka at speeds of around 135 mph (217 km/h). The word Shinkansen itself means "new main line" in Japanese but because of the trains' design and speed, they became known around the world as "bullet trains."

After the opening of the bullet trains in Japan, Europe also started developing high capacity high speed trains in 1965 at the International Transport Fair in Munich, Germany. Several high speed trains were tested at the fair but Europe's high speed rail service was not fully developed until the 1980s.

Today's High Speed Train Technology

Since the development of high speed rail, there have been many changes in the technology used in high speed trains. One of these is maglev(magnetic levitation)①, but most high speed trains use other technologies because they are easier to implement and they allow for more direct high speed connections to cities without the need for new tracks.

Today there are high speed trains that use steel wheels on steel tracks that can travel at speeds over 200 mph. Minimal stopping for traffic, long curves, and aerodynamic②, light trains also allow today's high speed trains to travel even faster. In addition, new technologies being implemented in train signaling systems can enable high speed trains to safely minimize time between trains at stations, thereby allowing travel on them to be even more efficient.

Worldwide High Speed Trains

Today, there are many large high speed rail lines around the world. The largest though are found in Europe, China and Japan. In Europe, high speed trains operate in Belgium, Finland, France, Germany, Italy, Portugal, Romania, Spain, Sweden, Turkey and the United Kingdom. Spain, Germany, the U.K. and France currently have the largest high speed train networks in Europe.

High speed trains are also significant in China and Japan. China, for example, has the world's largest high speed rail network at just over 3,728 miles (6,000 km). The network provides service between the country's major cities using maglev as well as more conventional trains.

Prior to China's construction of new high speed rail lines in 2007, Japan had the world's largest high speed train network at 1,528 mi (2,459 km). Today the Shinkansen is highly important there and new maglev and steel wheeled trains are currently being tested.

In addition to these three areas, high speed rail lines are also present as a commuter③ train in the eastern US and also in South Korea.

Advantages of High Speed Trains

Once completed and well established, high speed train lines have many advantages over other forms of high capacity public transportation. One of these is that due to infrastructure design in many countries, highway and air travel systems are constrained, cannot expand, and in many cases are overloaded. Because the addition of new high speed rail can also be high capacity, it has the potential relieve congestion on other transit systems.

High speed trains are also considered more energy efficient or equivalent to other modes of transit per passenger mile. In terms of possible passenger capacity, high speed trains can also reduce the amount of land used per passenger when compared to cars on roads. In addition, train stations are normally smaller than airports and can therefore be located within major cities and spaced closer together, allowing for more convenient travel.

Future of High Speed Trains

Because of these advantages, high speed rail use is increasing worldwide. By 2025 Europe plans to dramatically increase its connections and the EU has a goal of creating a Trans-European high speed train network to connect the entire region. Other examples of future high speed rail plans can be found across the globe from California to Morocco to Saudi Arabia, thus strengthening the importance of high speed trains as a viable form of future public transportation.

Notes:

① maglve 磁悬浮火车

② aerodynamic 空气动力学的

③ commuter 通勤者

Exercises

Work in groups and discuss the following questions.

1. How fast was the high speed train in 1962 in Japan?
2. Is maglve preferred in the modern high speed train technology? Why?
3. Which countries currently have the largest high speed train networks in Europe?
4. As a form of high capacity public transportation, what difference can the high speed trains make?
5. What are the differences between the high speed trains and the automobiles?

Unit 15 Digital Technology

Text A Digital Spies

In the final days of the Cold War, the crumbling Soviet Union possessed the nuclear weapons to destroy the world but lacked the economic and informational infrastructure to compete as a world power. While the preeminent① weapon for most of this century was the hydrogen bomb②, it has been replaced by the awesome capability of a single electron③! Future superpowers will be those nations with the greatest capability to harness the power of the electron for both economic and "digital" warfare.

The traditional Cold War alignment of the East vs. West is gone forever. Gen. Yuri Kobaladze of the SVR—the Russian Foreign Intelligence Service—recently stated, "there are friendly nations, but there are no friendly intelligence services".

At the height of Cold War solidarity (the enemy of my enemy is my friend), the major superpowers collected intelligence and attacked the ciphers④, or codes, of our friends as well as our enemies. The national interests of former friends and foes are now being redefined in terms of competing economic interests.

The traditional roles of spies in gathering, communicating and analyzing information (secrets), as well as counterintelligence, have been altered in ways never before imagined.

The advent of the Keyhole satellite program nearly 30 years ago provided the United States with the capability to digitally observe events on earth in near "real time". inclement⑤ weather and darkness. Using infrared⑥ cameras, radar and advanced sending lenses, they can resolve images approaching a single inch in diameter.

New "ears in space," sometimes officially designated as weather or mapping satellites, will continue to eavesdrop⑦ on all forms of communication signals transmitted into the ether. The increasing utilization of wireless frequencies for the transmission of telephone and computer data is absorbed into the antenna[8] of these satellites. Speech recognition software, new to the consumer market but utilized by intelligence for more than 25 years, will employ artificial intelligence to filter the unnecessary and recover secrets being communicated by both friends and foes.

The transformation of the Internet into the information highway has forever changed the way in which information is gathered. About 90 percent of everything spies need to know is available openly. The Internet, as the library of world knowledge, has become the repository of information needed fuel economies of the world's superpowers. The keys to this fountain of knowledge are high-speed Internet access, advanced networking to share information quickly, and massive computer power to analyze billions of bits of data to discover the secrets hidden inside. A clever computer programmer in the immediate future will recover more vital information in a day than a thousand fictional James Bonds could recover in a lifetime.

The most dangerous point of vulnerability for a spy operating in hostile territory was not when he was stealing secrets, but rather when he attempted to communicate them to his "handler". Public awareness of the "tradecraft" of the Cold War was often focused the communication techniques of "brush passes," sophistication and usefulness, they were vulnerable to an alert counterintelligence service and often confirmed the actions of the suspect being observed.

The Internet has changed this vulnerability into an advantage for the spy. Spies now utilize Internet to communicate with near impunity. Messages, information and signals are now transmitted in ways that appear innocuous[9] but almost defy detection because they are interlaced[10] into the normal and growing usage of the Internet. As information is transmitted or received into the Internet, its true recipient or sender may be masked in a bewildering variety of disguises. What once took days and weeks to communicate from a spy to his handler may now occur in milliseconds. Advanced encryption[11] techniques may be utilized to additionally mask data that may later be imbedded[12] into a digital scan, voice, music or television signal transmitted or received anywhere in the world. Even the world's most powerful computers lack the processing power to analyze trillions of bits of data for patterns to indicate possible imbedded messages.

In East Germany during height of the Cold War, a sea of information overwhelmed the human capabilities in place to transcribe and analyze the results. Even if a great secret had eventually been captured, the likelihood that it would be transcribed and analyzed in time to be useful was naught⑬. Without modern computers and the resulting analysis, the entire East German state eventually swamped⑭ itself in a sea of information.

The analysts have long been the unsung⑮ heroes of the spy world. With little fanfare they accumulate bits of information from sources around the world and convert them into a useful intelligence product. More powerful computers scan information from all sources to discern patterns and make predictions. The resulting analysis may be a weather pattern and resulting grain harvest in a foreign country predicted years in advance. Though apparently innocuous, such vital economic information becomes part of the finished intelligence product and potentially shapes foreign policy.

In the new world of the digital spy, computer viruses have been developed and deployed that will be activated in time of war. Imagine the consequence of embedding a "Trojan horse" in the operating system software that runs 90 percent of the computers of both friends and foes. A Trojan horse, once activated, can selectively disable the computer infrastructure of a hostile opponent and cripple its economy, communications and defense. The war is over before it has begun.

The traditional world of spies exists now only in fiction. Those intelligence services that most effectively identify, develop and implement the tools and techniques of the "cyber-spy" will provide their citizens with an incalculable advantage going into the new century.

Notes:

① preeminent 卓越的
② hydrogen bomb 氢弹
③ electron 电子
④ cipher 密码
⑤ inclement 险恶的
⑥ infrared 红外线的
⑦ eavesdrop 偷听
⑧ antenna 天线
⑨ innocuous 平淡无味的
⑩ interlaced 交织的
⑪ encryption 编密码
⑫ imbed 隐藏;埋入
⑬ naught 零
⑭ swamp 淹没
⑮ unsung 未被赞颂的

Exercises

Ⅰ. Read each of the following statements carefully and decide whether it is true or false according to the text.

1. The world's major intelligence agencies began to use the latest technology in collection, communication and analysis of information in World War I. (　　)
2. The national interests of former friends and foes are now being redefined in terms of competing economic interests. (　　)
3. The most dangerous thing for a spy operating in hostile territory was stealing secrets. (　　)
4. As information is transmitted or received into the Internet, its true recipient or sender may be masked in various disguises. (　　)
5. Modern computers play an important role in transcribing and analyzing information. (　　)
6. A Trojan horse, once activated, can completely destroy the computer infrastructure of a hostile opponent. (　　)

Ⅱ. Answer the following questions according to the text.

1. What does the case of the Soviet Union in the Cold War indicate?
2. How have the traditional roles of spies been altered?
3. What has forever changed the way in which information is gathered? And how?
4. What are the advantages of spies in the digital age?
5. How can computer viruses be used in the new world of the digital spy?

Ⅲ. Translate the following terms into their Chinese or English equivalents.

1. intelligence agency
2. counterintelligence agency
3. advanced sensing lenses
4. speech recognition software
5. encryption techniques
6. 数码扫描仪
7. 破译
8. 气象卫星
9. 人工智能
10. 信息高速公路

Ⅳ. Translate the paragraph into Chinese.

The transformation of the Internet into the information highway has forever changed the way in which information is gathered. About 90 percent of everything spies need to know is available openly. The Internet, as the library of world knowledge, has become the repository of information needed fuel economies of the world's superpowers. The keys to this fountain of knowledge are high-speed Internet access, advanced networking to share information quickly, and massive computer power to analyze billions of bits of data to discover the secrets hidden inside. A clever computer programmer in the immediate future will recover more vital information in a day than a thousand fictional James Bonds could recover in a lifetime.

Text B How to Take Advantage of Mobile Apps for Monitoring Your Health

Many people think that health care is what happens at their health provider's office, with health monitoring limited to time spent at a clinic or a medical facility. However, in reality managing one's health is ever-present once we are old enough to become self-aware. However, and fortunately, most people spend the majority of their lives away from medical settings. Most would agree that health is not top of mind while engaging in day-to-day activities.

Nonetheless, good health is an ongoing pursuit—a value that ideally should permeate① our motives and actions. Unfortunately, our personal well-being is often not contemplated until an ill-fated event makes us mindful of the importance of health.

In contrast to healthy individuals, people who suffer from chronic diseases are quite aware of the need for perpetual health tracking. Measuring vital signs and regularly observing their body's biometric signals are often an integral② part of their daily routines as well as a life-long commitment in their treatment regimen③.

New technological developments are offering hope that these individuals can become more connected, making the data they collect more available, valuable and useful.

Retrofitting Antiquated④ Technology

One of the key barriers to device interoperability⑤ is the speed at which health technology is developed. Many devices that were created only a few years back were not built with transmitting data in mind, let alone with the thought of data standards.

One of the first companies to try to solve this problem is Netpulse. Netpulse, a fitness technology company, was an early pioneer with the idea of integrating data through image capture using xCapture feature from their smartphone application. Predominantly promoted in the health club space, the Netpulse mobile application aims to digitalize fitness data from disparate exercise equipment through pattern recognition and then coalesce⑥ this information so it becomes more useful for the user.

The Netpulse app gives users the freedom to work out on a variety of equipment without being confined to a particular brand's proprietary⑦ front-end software.

Depositing Health Data onto a Phone

Evolving this concept for health care, Validic, one of the world's leading digital health platforms, is introducing mobile health technology that can transfer real-time data from Non-Connected Devices to Health IT Systems. Announced at this year's CES, Validic's VitalSnap can help patients reach their health goals and could potentially revolutionize the way health data is collected and used. Drew Schiller, Validic's co-founder and Chief Technology Officer, recognized that a majority of medical devices people use today to monitor their personal health needs are not connected and therefore do not sync⑧ with health data platforms. These so-called Non-Connected Devices limit the utility of data that patients with chronic conditions gather during daily self-monitoring.

Information collected with many early monitoring devices often leaves the user relying on pen and paper to track progress.

Manual data collection methods have a lot of limitations in terms of storing and sharing information in a meaningful way. Moreover, they are often not practical nor are they compatible with modern data collection practices—let alone the tempo of modern life.

The value of VitalSnap is it allows patients to continue using existing non-connected health devices by adding the ability to connect their data in a novel way. By simply using a smartphone, a patient can now store his or her readings onto the accompanying application. After taking his or her biometric measurement, the user holds the smartphone over the device and opens the VitalSnap application. The device's readings get instantly captured by the phone's camera. The data are then delivered to the system via the Validic Digital Health Platform and can be sent directly to the individual's health-care provider as well, maintaining privacy and security at all times. This enables patients to participate in their health care in a way that had not been possible before.

Tracking health becomes not only more effective but possibly more enjoyable and compatible with other daily activities. Also, this innovative digital technology is aiming to bridge the gap between patients who have difficulty traveling but still would like to share data with their health-care provider. When it comes to data interoperability health care lags behind many industries, but VitalSnap is enabling patients to meet their providers where they are today creating a win-win situation.

Notes:

① permeate 渗透
② integral 不可或缺的
③ regimen 养生之道
④ antiquated 过时的
⑤ interoperability 互通(性);互用(性)
⑥ coalesce 结合;合并
⑦ proprietary 专营的;专利的
⑧ sync 同步

Exercises

Work in groups and discuss the following questions.

1. What kind of problem does Netpulse Co. try to solve?
2. What is the purpose of Netpulse mobile application?
3. How do you define the mobile health technology introduced by Validic?
4. What are the disadvantages of manual data collection methods?
5. What makes digital health-tracking technology prevalent among patients?

Unit 16 Ecology

Text A Ecology of Transgenic Crops

On May 20, 1999, a short article in *Nature* called attention to a potential ecological problem with a genetically engineered, or transgenic crop①. John Losey and his colleagues at Cornell University reported that a variety of transgenic corn could kill the larvae② of monarch butterflies. Opponents of transgenic crops held up the report as evidence of the potentially devastating environmental impact of this new technology. Proponents, on the other hand, largely dismissed this laboratory-based research as unrepresentative of conditions on real farms. Yet despite the disagreements, this study draws attention to the difficulty of determining the safety or danger of this new generation of crops.

Genetic engineering makes it possible to transfer genes from virtually any species—animal, bacteria, plant or virus—into almost any other species, no matter how unrelated the two species might be. Consequently, these revolutionary molecular techniques let scientists generate organisms with entirely new combinations of properties. For example, a jellyfish gene transferred to plants makes them luminescent③, and the Monsanto Corporation is developing new varieties of grass that will produce colored lawns.

Beyond these more fantastic applications, genetically engineered crops might offer valuable benefits: increased yields, improved flavor or nutritional quality of foods and reduced pesticide use. On the other hand, transgenic crops also pose potential risks. Most public attention has focused on harmful effects to human health, including the production of novel allergens④ or carcinogens⑤. But there is also a range of possible environmental impacts, including increased reliance on herbicides⑥, the creation of new pests, harmful effects on non-target species and the disruption of ecosystem processes—concerns that have been the focus of my work.

Unfortunately, scientists lack the necessary data to predict the consequences of widespread commercial planting of transgenic crops, largely because the technology itself remains so new. Nonetheless, transgenic crops are currently being planted on a commercial scale, and the area devoted to transgenic crops increased from 4.3 million acres in 1996 to 69.5 million acres in 1998. With such rapidly increasing use of transgenic crops, scientists and society must weigh whether the potential benefits outweigh the potential risks.

Scientists must ask: Do transgenic crops pose different risks from those common to crops created through traditional methods of plant breeding? After all, plant breeders used traditional methods for millennia to create organisms with quite novel traits. For example, varieties as distinct as broccoli, Brussels sprouts and cabbage came from a single species of mustard. Many scientists emphasize that the product—not the process—needs to be regulated and evaluated for risk. In other words, transgenic crops should not require regulation simply because they are genetically engineered. Instead, a transgenic crop should be regulated only if it is likely to pose elevated threats to human health or the environment. Nevertheless, genetic engineering can create many more combinations of genes and new traits than can traditional breeding. This greatly enhanced novelty diminishes anyone's ability to predict the safety of a transgenic organism on the basis of past experience.

Proponents of genetic engineering call attention to the lack of major trouble associated with transgenic crops, but that record does not guarantee that they are completely safe. As a case in point, scientists and manufacturers considered pesticides totally risk-free when first marketed in the late 1940s, and data that documented ill effects took nearly 20 years to surface. Similarly, major problems might result from transgenic crops over time. So far, few experiments have examined the safety of transgenic crops, especially the many ways that these modifications could affect the environment. Moreover, the companies that manufacture and market transgenic crops are responsible for assessing their safety, which results in a potential conflict of interest that could compromise the rigor of safety assessments. Finally, because some of the risks derive from rare, chance events—including hybridization⑦ between a crop and a weedy relative—it might take quite some time for troubles to emerge. Meanwhile, money and time for monitoring the environment after release of transgenic crops are limited, making it likely that no one would detect any sign of trouble until well after a problem developed.

To illustrate the challenges associated with assessing the risks and merits of this new technology, I shall focus on insect-resistant transgenic crops. The safety of crops engineered to be toxic or repellent to insects is a crucial concern, because such crops made up about one-quarter of the total cotton acreage and one-fifth of the total corn acreage in the United States in 1998. This rapidly increasing use of transgenic crops demands heightened rigor in testing these novel plants.

Creating New Weeds

The release of organisms with novel phenotypes[⑧] bears similarities to the introduction of non-native species. Many well-documented examples reveal non-native plants, including kudzu and purple loosestrife, becoming aggressive weeds with devastating environmental and economic consequences. Sometimes, introduced plants invade successfully because no insect herbivores attack them. Consequently, insect-resistant transgenic plants might be more likely to become invasive weeds than would the parental variety.

Moreover, hybridization between a transgenic crop and a related noncrop plant can spread novel traits to additional species, which further complicates the analysis of the risk of creating new weeds. For example, Norman Ellstrand and his colleagues at the University of California at Riverside found that 12 of the world's 13 most important food crops hybridize with wild relatives in some part of their distributions. If a transgenic crop can hybridize with nearby wild relatives, the transfer of genes will be virtually inevitable once farmers plant the crop on a commercial scale. As a result, insect-resistance traits could create aggressive weeds from either the crop or, more likely, from related noncrop species. Still, some scientists argue that the escape of insect-resistance genes into wild populations would not substantially increase the population growth rate of wild plants.

Without direct field results indicating how herbivore-resistant crops could generate weeds, some insight can be gained from ecological experiments on herbivory in natural plant populations. To quantify the impact of herbivores, scientists often manipulate the number of herbivores attacking plants and then measure changes in plant growth or reproduction. To speculate on the future impacts of transgenic crops, an investigator must summarize a large collection of such studies across many species and situations, which can be done with a statistical approach called meta-analysis. Peter Kareiva of the Marine Fisheries Service and I applied this method to results reported in 18 different publications that involved 52 different plant-herbivore combinations. We found an extraordinarily large average effect of herbivory: Plants protected against invertebrate[⑨]—primarily insect—herbivores produced more reproductive structures than did 81 percent of unprotected plants. Thus any herbivore-resistance trait is likely to confer a substantial advantage, which could easily increase the occurrence of weeds. Although increases in seed production do not always translate into enhanced weediness, limitations on seed production do constrain many natural plant populations, at least in some years. Consequently, insect-resistance genes could cause nondomesticated relatives of transgenic crops to become weeds.

The analogy between the commercial-scale planting of transgenic crops and the introduction of non-native plant species suggests several additional factors to consider when assessing the risks associated with transgenic crops. First, the vast majority of introduced plant species cause no serious environmental problems. Accordingly, most transgenic crops will probably pose little threat to the environment. As in plant introductions, though, a small percentage of transgenic varieties might become serious pests that cause vast economic and environmental damage. Second, even introduced plants that become aggressive weeds often remain relatively uncommon for long periods before becoming problematic. For example, *Mimosa pigra*[10], or catclaw mimosa[11], was a minor weed in Australia for about a century before it expanded dramatically and excluded other plants over large areas. In addition, my colleagues and I searched historical reconstructions of plant invasions and found many examples of remarkably long lags between the time of a plant's introduction and the detection of weed spread. Although there would be enormous pressure to discontinue monitoring efforts of transgenic crops if weeds have not been detected after many years, we must remember that detection lags can be quite long and that effective monitoring might require 30 or more years of sampling.

Dealing with Uncertainty

The public knows that some past technologies—including DDT, PCBs and others—caused major environmental problems, despite repeated assurances of safety from scientists. Undoubtedly, that history contributed to the public's current concern over transgenic crops. Those who stand to profit from transgenic technologies view this heightened public vigilance as unfounded, even hysterical, and caution that increased regulatory oversight would hinder progress. Nevertheless, the public stands to lose if transgenic crops cause damage to the environment.

Risk analysis should reveal how the public good might suffer if new technologies backfire. Nevertheless, transgenic crops offer some special challenges in applying ecological-risk analysis. Currently, investigators evaluate risks by analogy or by direct experiments. Analogies to exotic, invasive plants fall short, because transgenic crops are not entirely new to an area, but rather are modifications involving only a few traits. Analogies to familiar crops also fail, because the mingling of genes in transgenic crops can produce completely novel traits. This leaves direct experimentation and monitoring as the primary tools for risk assessment. To be completely assured of the ecological safety of a transgenic crop, however, would require many experiments: testing the transgenic plant in different environmental conditions, at different times of year, in combination with different farming practices, and examining the effects of the plants and plant by-products on an enormous number of species that could potentially be affected by the transgenic traits. Clearly, attaining this level of certainty is neither reasonable nor possible.

Consequently, we must decide how we will deal with the unavoidable uncertainty that accompanies transgenic crops. In essence, current regulations assume that a transgenic crop is safe unless it is shown otherwise. Alternatively, we could assume that a transgenic product is unsafe until the manufacturer demonstrates its safety. Despite recent studies that highlight possible risks, plants engineered to express Bttoxin⑫ are almost certainly safer than most chemical pesticides, which generate well-established dangers for nontarget arthropods. Nevertheless, an extraordinary variety of novel traits will be incorporated into tomorrow's transgenic plants, and the safety of those plants will vary substantially. If the assumption remains "safe until proven otherwise," regulators must boost the rigor of testing. Experiments with a handful of replicates and low statistical power could fail to expose the environmental risks associated with particular transgenic varieties.

Notes:

① transgenic crop 转基因作物

② larvae (sing.) larva 幼虫

③ luminescent 发冷光的

④ allergen 变应原

⑤ carcinogen 致癌物

⑥ herbicide 除草剂

⑦ hybridization 杂交;配种

⑧ phenotype 表型

⑨ invertebrate 无脊椎的

⑩ *Mimosa pigra* 刺轴含羞草,是一种多刺的多年生草本或亚灌木,属于豆科含羞草属,该物种原产美洲,后入侵台湾

⑪ catclaw mimosa 含羞草

⑫ Bttoxin 一种毒药

Exercises

Ⅰ. Read each of the following statements carefully and decide whether it is true or false according to the text.

1. Genetic engineering makes it possible to transfer genes from virtually any species into almost any other species. ( )
2. Scientists can predict the consequences of widespread commercial planting of transgenic crops. ( )
3. Transgenic crops should be regulated simply because they are genetically engineered. ( )
4. Major problems may result from transgenic crops over time. ( )
5. Current regulations assume that a transgenic crop is safe unless it is shown unsafe. ( )
6. Experiments with a handful of replicates can reveal the environment risks associated with some transgenic crops. ( )

Ⅱ. Answer the following questions according to the text.

1. What are the advantages of genetically engineered crops?
2. What potential risks do transgenic crops pose?
3. How to speculate on the future impacts of transgenic crops?
4. Is it possible to be completely assured of the ecological safety of a transgenic crop? Why ?
5. What attitude should we take towards transgenic crops?

Ⅲ. Translate the following terms into their Chinese or English equivalents.

1. commercial planting
2. genetic engineering
3. pesticide
4. herbivore-resistance
5. introduced plants
6. 转基因作物
7. 基因
8. 细菌
9. 复制品
10. 杂交

Ⅳ. Translate the paragraph into Chinese.

Similarly, major problems might result from transgenic crops over time. So far, few experiments have examined the safety of transgenic crops, especially the many ways that these modifications could affect the environment. Moreover, the companies that manufacture and market transgenic crops are responsible for assessing their safety, which results in a potential conflict of interest that could compromise the rigor of safety assessments. Finally, because some of the risks derive from rare, chance events—including hybridization⑦ between a crop and a weedy relative—it might take quite some time for troubles to emerge. Meanwhile, money and time for monitoring the environment after release of transgenic crops are limited, making it likely that no one would detect any sign of trouble until well after a problem developed.

Text B Keystone Species and Their Role in Ecology

Keystone Species Definition

A keystone species is a plant or animal that plays a vital role in the health of the ecosystem in which it lives.

To understand the role of a keystone species, think about the role of a keystone in an arch. An arch is built up of many bricks. Take away a brick along its side or in a corner and the arch may wobble, but it will likely still stand. But take away the keystone—the vital brick in the center of the arch, and the entire structure will collapse.

The same can be said for keystone species. While all plants and animals play an important role in the health of an ecosystem, it is the vitality of its keystone species that determines whether or not the ecosystem will survive.

History of the Keystone Species Concept

The term "keystone species", was first used in 1969 by University of Washington Zoology professor Robert T. Paine. In his studies of the Makah Bay① in Washington, Paine described a species of starfish, *Pisaster ochraceus*②, as a primary example of a keystone species whose population health could determine the overall health of the aquatic ecosystem within the bay.

Examples of Keystone Species

Because the definition of the concept is broad, ecologists don't always agree on which species within an ecosystem are the keystone species. So rather than look for specific examples of keystone species, it's better to understand the characteristics that define them.

Predators[3] as Keystone Species

More often than not, a keystone species is a predator. One predator species can control the population size of several smaller animals.

For example, mountain lions are considered keystone species, as their numbers directly influence the population size of rabbits, deer, birds, and even scavengers[4] such as vultures[5] that live within the same habitat.

The key to understanding the role of the keystone species is understanding the effect its disappearance would have on the health of the ecosystem.

Using the mountain lion example mentioned above, if mountain lions suddenly went extinct, it's safe to assume that the populations of deer and rabbits would explode and then crash as the overpopulations of the species competed with one another for food. Grasses, shrubs, and flowers would quickly become depleted. And while the initial death toll of animals would be a boon for the scavengers, the supply would dry up as the populations crash, leaving the vultures with little left to scavenge.

Mutualists Keystone Species[6]

Keystone species are often—but not always—predators. Sometimes, the keystone species is the one whose actions benefit others in the ecosystem in such a way that without them, other species would not survive.

Certain species of hummingbirds[7] in the Sonoran Desert of the southwest US are considered keystone species as they are the only birds that pollinate numerous local plant species within the environment. In areas where the hummingbird populations have declined, non-native species have taken over and native plant species are starting to disappear as a result.

Other examples of keystone species include the sea otter of the Pacific Northwest, elephants of the African savannah, prairie dogs, parrotfish, and beavers.

Importance of Keystone Species

The health of a keystone species population is directly related to the health of the entire ecosystem. While other species may come and go, the loss of a keystone would make it impossible for many of the habitats other species to survive.

Notes:

① the Makah Bay 马考海湾

② *Pisaster ochraceus* 紫海星

③ predator 食肉动物

④ scavenger 食腐动物

⑤ vulture 兀鹫;秃鹫

⑥ mutualists keystone species 关键互惠共生种

⑦ hummingbird 蜂鸟

Exercises

Work in groups and discuss the following questions.

1. What is a keystone species?
2. What is the origin of the term "keystone species"?
3. How to understand the role of predators in ecosystem?
4. What is mutualists keystone species?
5. What does the loss of a keystone bring about in ecosystem?

Unit 17 Oceanography

Text A An Introduction to Oceanography

Oceanography is a discipline within the field of Earth sciences (like geography) that is focused entirely on the ocean. Since the oceans are vast and there are many different things to study within them, the topics within oceanography vary but include such things as marine organisms and their ecosystems, ocean currents, waves, seafloor geology (plate tectonics included), the chemicals making up seawater and other physical characteristics within the world's oceans.

In addition to these broad topic areas, oceanography includes topics from a number of other disciplines like geography, biology, chemistry, geology, meteorology and physics.

History of Oceanography

The world's oceans have long been a source of interest for humans and people first began gathering information about waves and currents hundreds of years ago. Some of the first studies on tides were collected by the Greek philosopher Aristotle and the Greek geographer Strabo.

Some of the earliest oceanic explorations were in an attempt to map the world's oceans to make navigation easier. However, this was mainly limited to areas that were regularly fished and well-known. This changed in the 1700s though when explorers like Captain James Cook extended their explorations into previously unexplored regions. During Cook's voyages from 1768 to 1779 for example, he circumnavigated areas such as New Zealand, mapped coastlines, explored the Great Barrier Reef① and even studied portions of the Southern Ocean.

During the late 18th and into the early 19th centuries, some of the first oceanographic textbooks were written by James Rennell, an English geographer and historian, about ocean currents. Charles Darwin also contributed to the development of oceanography in the late 1800s when he published a paper on coral reefs and the formation of atolls after his second voyage on the HMS Beagle②.

The first official textbook covering the various topics within oceanography was later written in 1855 when Matthew Fontaine Murray, an American oceanographer, meteorologist and cartographer, wrote Physical Geography of the Sea.

Shortly thereafter, oceanographic studies exploded when the British, American and other European governments sponsored expeditions and scientific studies of the world's oceans.

These expeditions brought back information on ocean biology, physical formations and meteorology.

In addition to such expeditions, many oceanographic institutes were formed in the late 1880s. For example, the Scripps Institution of Oceanography was formed in 1892. 1902, the International Council for the Exploration of the Sea was formed; creating the first international organization of oceanography and in the mid-1900s, other research institutions focused on oceanography were formed.

Recent oceanographic studies have involved the use of modern technology to gain a more in depth understanding of the world's oceans. Since the 1970s for example, oceanography has emphasized the use of computers to predict ocean conditions. Today, studies focus mainly on environmental changes, climate phenomena like El Niño and sea floor mapping.

Topics in Oceanography

Like geography, oceanography is multi-disciplinary and incorporates a number of different sub-categories or topics. Biological oceanography is one of these and it studies the different species, their living patterns and interactions within the sea. For example, different ecosystems and their characteristics such as coral reefs versus kelp forests③ can be studied within this topic area.

Chemical oceanography studies the different chemical elements present in seawater and how they interact with the Earth's atmosphere. For example, nearly every element in the periodic table is found in the ocean. This is important because the world's oceans serve as a reservoir for elements like carbon, nitrogen and phosphorus—each of which can impact the Earth's atmosphere.

Ocean/atmosphere interactions is another topic area in oceanography that studies the links between climate changes, global warming and concerns for the biosphere as a result. Mainly, the atmosphere and oceans are linked because of evaporation and precipitation. In addition, weather patterns like wind drive ocean currents and move around different species and pollution.

Finally, geological oceanography studies the geology of the seafloor (such as ridges and trenches) and plate tectonics, while physical oceanography studies the ocean's physical characteristics which include the temperature-salinity structure, mixing levels, waves, tides and currents.

Importance of Oceanography

Today, oceanography is a significant field of study throughout the world. As such, there are many different institutions devoted to studying the discipline such as the Scripps Institution of Oceanography, The Woods Hole Oceanographic Institution and the United Kingdom's National Oceanography Centre in Southampton. Oceanography is an independent discipline in academic with graduate and undergraduate degrees being issued in oceanography.

In addition, oceanography is significant to geography because the fields have overlapped in terms of navigation, mapping and the physical and biological study of Earth's environment—in this case the oceans.

Interesting Oceanography Facts

- It is believed that approximately 95% of the oceans are yet unexplored. Scientists know more about the moon and Mars then they do about the ocean.
- The first person to study the Gulf Stream scientifically was Benjamin Franklin.
- There are hundreds of earthquakes around the world every day but most are under the water of the oceans.
- Some of these earthquakes cause tsunamis which are giant underwater waves. If these waves reach land they can result in massive destruction and death.
- The water in the ocean is not drinkable because it is salt water.
- The Mid-Ocean Ridge is the longest mountain range in the world and it is located under the ocean. It is more than 45,000 miles long and winds around the globe.
- The first textbook about oceanography was called the *Physical Geography of the Sea*. It was written by Matthew Fontaine Maury in 1855.
- The country with the longest ocean coastline is Canada, with 56,453 miles of coastline length.
- There is gold suspended in seawater. Mining it from the seawater has not been possible

as of yet. If it was possible it is estimated that every person on earth could have nine pounds of gold.

- The Antarctica is melting due to global warming. This is resulting in rising sea levels. If global warming does not stop, the melting glaciers will result in flooding in cities that are on the coasts and that are not high enough above sea levels. Some of these cities include New York, London, and Mumbai.
- Oceanographers believe that the ocean floors are only about 200 million years old while the continents are about 2–3 billion years old.
- The highest ocean tide in the world is in the Bay of Fundy on Canada's east coast. It can range as much as 53.5 feet in the spring.
- Ocean tides in Alaska can range as much as 40 feet.
- Ocean waves can be caused by the wind, by earthquakes, and by other underwater phenomena.
- The largest fish in the oceans, and in the world, is the whale shark which can grow longer than 40 feet.
- The smallest fish in the world is the dwarf goby which is only 0.3 inches long.
- The oceans' coral reefs are ranked second in terms of biodiversity of species. The world's rainforests rank first.

Notes:

① the Great Barrier Reef 大堡礁，位于澳大利亚东北部，由约3000个不同阶段的珊瑚礁、珊瑚岛、沙洲和潟湖组成，是世界最大最长的珊瑚礁群

② HMS Beagle 英国皇家海军"小猎犬号"或贝格尔号

③ kelp forest 海藻林

Exercises

Ⅰ. Read each of the following statements carefully and decide whether it is true or false according to the text.

1. Oceanographic studies originated from gathering information about waves and currents. ( )
2. Physical oceanography studies the ocean's geological characteristics. ( )
3. Students of oceanography can be issued graduate and undergraduate degrees. ( )
4. The first textbook about ocean currents was written by Matthew Fontaine Murray. ( )
5. Oceanography includes a rich variety of topics. ( )

6. Tsunamis are giant underwater waves. ( )

Ⅱ. **Answer the following questions according to the text.**

1. What is oceanography?
2. What are the recent oceanographic studies like?
3. What are the topics in oceanography?
4. Is oceanography important to geography? Why?
5. How do ocean waves come about?

Ⅲ. **Translate the following terms into their Chinese or English equivalents.**

1. oceanographer
2. the Great Barrier Reef
3. ocean currents
4. sea floor mapping
5. geological oceanography
6. 气象学
7. 生物海洋学
8. 海藻林
9. 珊瑚礁
10. 海啸

Ⅳ. **Translate the paragraph into Chinese.**

Ocean/atmosphere interactions is another topic area in oceanography that studies the links between climate changes, global warming and concerns for the biosphere as a result. Mainly, the atmosphere and oceans are linked because of evaporation and precipitation. In addition, weather patterns like wind drive ocean currents and move around different species and pollution.

Text B Ocean Waves

Waves are the forward movement of the ocean's water due to the oscillation of water particles by the frictional drag of wind over the water's surface.

Size of a Wave

Waves have crests (the peak of the wave) and troughs (the lowest point on the wave). The wavelength, or horizontal size of the wave, is determined by the horizontal distance between two crests or two troughs. The vertical size of the wave is determined by the vertical distance between the two.

Waves travel in groups called wave trains.

Different Kinds of Waves

Waves can vary in size and strength based on wind speed and friction on the water's surface or outside factors such as boats. The small wave trains created by a boat's movement on the water are called wake①. By contrast, high winds and storms can generate large groups of wave trains with enormous energy.

In addition, undersea earthquakes or other sharp motions in the seafloor can sometimes generate enormous waves, called tsunamis (inappropriately known as tidal waves) that can devastate entire coastlines.

Finally, regular patterns of smooth, rounded waves in the open ocean are called swells②. Swells are defined as mature undulations of water in the open ocean after wave energy has left the wave generating region. Like other waves, swells can range in size from small ripples to large, flat-crested waves.

Wave Energy and Movement

When studying waves, it is important to note that while it appears the water is moving forward, only a small amount of water is actually moving. Instead, it is the wave's energy that is moving and since water is a flexible medium for energy transfer, it looks like the water itself is moving.

In the open ocean, the friction moving the waves generates energy within the water. This energy is then passed between water molecules in ripples called waves of transition. When the water molecules receive the energy, they move forward slightly and form a circular pattern.

As the water's energy moves forward toward the shore and the depth decreases, the diameter of these circular patterns also decreases.

When the diameter decreases, the patterns become elliptical and the entire wave's speed slows. Because waves move in groups, they continue arriving behind the first and all of the waves are forced closer together since they are now moving slower. They then grow in height and steepness. When the waves become too high relative to the water's depth, the wave's stability is undermined and the entire wave topples onto the beach forming a breaker③.

Breakers come in different types—all of which are determined by the slope of the shoreline. Plunging breakers are caused by a steep bottom; and spilling breakers signify that the shoreline has a gentle, gradual slope.

The exchange of energy between water molecules also makes the ocean crisscrossed with waves traveling in all directions. At times, these waves meet and their interaction is called interference, of which there are two types. The first occurs when the crests and troughs between two waves align and they combine. This causes a dramatic increase in wave height. Waves can also cancel each other out though when a crest meets a trough or vice versa. Eventually, these waves do reach the beach and the differing size of breakers hitting the beach is caused by interference farther out in the ocean.

Ocean Waves and the Coast

Since ocean waves are one of the most powerful natural phenomena on Earth, they have a significant impact on the shape of the Earth's coastlines. Generally, they straighten coastlines. Sometimes though, headlands④ composed of rocks resistant to erosion jut into the ocean and force waves to bend around them. When this happens, the wave's energy is spread out over multiple areas and different sections of the coastline receive different amounts of energy and are thus shaped differently by waves.

One of the most famous examples of ocean waves impacting the coastline is that of the longshore or littoral current. These are ocean currents created by waves that are refracted as they reach the shoreline. They are generated in the surf zone when the front end of the wave is pushed onshore and slows. The back of the wave, which is still in deeper water moves faster and flows parallel to the coast. As more water arrives, a new portion of the current is pushed onshore, creating a zigzag pattern in the direction of the waves coming in.

Longshore currents are important to the shape of the coastline because they exist in the surf zone and work with waves hitting the shore. As such, they receive large amounts of sand and other sediment and transport it down shore as they flow. This material is called longshore drift⑤ and is essential to the building up of many of the world's beaches.

The movement of sand, gravel and sediment with longshore drift is known as deposition. This is just one type of deposition affecting the world's coasts though, and have features formed entirely through this process. Depositional coastlines are found along areas with gentle relief and a lot of available sediment.

Coastal landforms caused by deposition include barrier spits⑥, bay barriers⑦, lagoons⑧, tombolos⑨ and even beaches themselves. A barrier spit is a landform made up of material deposited in a long ridge extending away from the coast. These partially block the mouth of a bay, but if they continue to grow and cut off the bay from the ocean, it becomes a bay barrier. A lagoon is the water body that is cut off from the ocean by the barrier. A tombolo is the landform created when deposition connects the shoreline with islands or other features.

In addition to deposition, erosion also creates many of the coastal features found today. Some of these include cliffs, wave-cut platforms, sea caves and arches. Erosion can also act in removing sand and sediment from beaches, especially on those that have heavy wave action.

These features make it clear that ocean waves have a tremendous impact on the shape of the Earth's coastlines. Their ability to erode rock and carry material away also exhibits their power and begins to explain why they are an important component to the study of physical geography.

Notes:

① wake 尾波

② swell 涌浪;巨浪

③ breaker (尤指冲到海岸的)大浪

④ headland 陆岬; 海角

⑤ longshore drift 沿岸漂砂;沿岸飘移

⑥ barrier spit 障壁沙嘴;堰洲嘴

⑦ bay barrier 湾口沙坝;海湾口障壁坝

⑧ lagoon 潟湖

⑨ tombolo 沙颈岬;陆连岛; 连岛沙洲

Exercises

Work in groups and discuss the following questions.

1. What are waves?
2. How to determine the size of a wave?
3. How many different kinds of waves and what are they?
4. What's a breaker?
5. How do ocean waves impact the shape of the Earth's coastlines?

Unit 18 Environmental Technology

Text A How to Stop Humans from Filling the World with Trash

When the $ 20 Billion Hudson Yards development is finished on Manhattan's Far West Side in 2024, it will have six skyscrapers, 5,000 apartments, more than 100 stores, and a public school. One thing it will not have is municipal garbage trucks. Related Companies, one of the developers working on the project, plans to install pneumatic tubes① that will whisk trash to a sorting area. The system should decrease the amount of garbage that ends up in landfills: residents will be able to drop recyclables and compost into designated chutes right outside their doors. By replacing trucks, the tubes will also cut down on noise and pollution—and, hopefully, on rats.

New York has experimented with pneumatic tubes before—the city's Roosevelt Island has used them for trash since the 1970s—but they may become more common as cities struggle with the acres of trash their residents create.

The average American produces about 130 pounds of trash a month, and an article in the journal *Nature* estimates that global solid-waste generation will triple, to 11 million tons a day, by 2100. Meanwhile, we're running out of space for landfills, especially in Japan and Europe. Here, drawn from interviews with scientists, environmentalists, and sanitation experts, are ideas for how to tackle this looming problem.

Putting a Price on Trash

One way to get people to produce less garbage is to charge them for it. So-called pay-as-you-throw programs—in which municipalities bill residents for their garbage—have been around for decades but are becoming more widespread. And they work: since beginning a pay-as-you-throw progam in 1993, Worcester, Massachusetts, has seen a 53 percent drop in waste, from 43,000 tons a year to 20,000. "It really does change behavior," says Mark Dancy, the president of WasteZero, a company that runs similar programs in hundreds of municipalities. "Now that people are aware that trash has a cost, they begin looking for all the alternatives to putting things in a trash can."

Of course, people will try to cheat by putting their garbage in someone else's can. (Some towns employ inspectors to follow up on reports of illicit dumping; they search trash for identifying information and write tickets.) But technology could combat this, says Bryan Staley, the president of the Environmental Research and Education Foundation, which funds research on waste management. Companies could attach radio-frequency identification (RFID) tags—which cost as little as 7 cents a piece—on trash bags, he says, so that an RFID reader on a truck could reject any bags that don't belong to that household. RFID readers can also reward good behavior—a New Jersey trash-collection company called Sanico uses RFID chips on recycling bins to give people discounts when they recycle.

A Single Stream

Americans are pretty bad at composting and recycling: by some estimates, up to three-quarters of the materials in US landfills could have been diverted. Some experts think we should just collect everything —glass, paper, half-eaten Twinkies—in one bin and leave the messy work of sorting to robots. Certain municipalities that no longer require residents to separate paper from plastic and so forth already use machines to do much of this work. Eventually, Staley says, technology akin to facial-recognition software could further automate sorting by helping machines distinguish, say, a peanut-butter sandwich from a peanut-butter jar and send them along for composting and recycling, respectively. "You in essence remove this element of human behavior that requires people to make a decision about whether to throw something in the bin," Staley says.

Skyscrapers Made of Garbage

We already turn water bottles into fleece, plastic bags into deck material, roofing into pavement. But ideas abound for more-futuristic forms of recycling. Mitchell Joachim, a co-founder of Terreform ONE, a design firm based in New York, proposes crushing trash and molding it into Tetris esque blocks that we could use to build islands and skyscrapers. Joachim's firm has created architectural plans for a 53-story tower made with the waste New Yorkers produce in Guatemala called Pura Vida is already working on a low-tech version of the same idea; it promotes the use of a building material it calls an "echo-block" —just a plastic bottle stuffed with trash—that it says makes for excellent insulation and is safe in earthquakes.

Smarter Leftovers

Food accounts for about one-fifth of what goes into municipal landfills, and companies are looking for new ways to repurpose what we don't eat. Some farmers use leftovers to feed their animals, and companies in California and Ireland are turning edible trash into pet food. Better systems to collect and distribute excess food from grocery stores and restaurants could help feed the hungry. Such food recycling is difficult and labor-intensive because it has to be done very quickly, but as droughts challenge agricultural production around the world, it could become more common.

Garbage Power

Food can also be turned into fuel through anaerobic digestion②, a natural process during which microbes③ break down organic matter in the absence of oxygen. Farmers have used this process for years to make biogas out of manure; now new machines can speed things up. Anaerobic-digestion facilities are expensive to build, but they can be profitable if companies have a steady supply of food waste, as they would in the growing number of cities and states that have banned restaurants and grocery stores from sending large amounts of leftovers to landfills. Someday, business could build their own digesters, says Thomas DiStefano, a civil and environmental engineer at Bucknell University. Michigan State University has two such digesters that turn food waste from dining halls into electricity for the campus.

Burning garbage is another way to turn trash into fuel (usually by making steam, which turns turbines). The first trash incinerator④ in the US was built in 1885, and until the 1980s, we burned much of the waste we couldn't (or didn't) recycle. But scientists discovered that dioxin emissions from incineration plants caused cancer and birth defects. The technology has since improved, and today's plants are so clean that in Europe, builders are putting them in the middle of cities so they can power nearby households. In Copenhagen, a ski slope will be built atop one.

Plasma gasification⑤, an experimental technique, could eventually replace incineration as an even cleaner and more efficient way to get rid of trash, says Juliette Spertus, an architect who has studied waste management. The process involves heating waste under pressure to produce syngas, a substance that can be used to make liquid fuels and other chemicals. Another process, called pyrolysis⑥, also uses heat to turn trash into fuel. Both techniques are currently expensive and can process only small amounts of waste at a time, but they could become viable as space in landfills becomes increasingly scarce.

Ending Trash for Good

If rocket technologies improve, Staley says, we might one day blast trash into space and use the sun's heat to burn it. But given that our planet has limited resources, burning them after one use probably isn't the answer. Some environmentalists want to prevent companies from making nonrecyclable materials in the first place, and a few have suggested alternatives. A European research group called ZeroWIN, for example, designed a laptop made of recycled materials whose components can be reused. (Most computers end up in landfills, potentially leaking chemicals into the ground.) Joachim, of Terreform ONE, says the planned obsolescence⑦ of products should be outlawed. So-called extended-producer-responsibility laws could require manufacturers to fund and manage the recycling of their goods so that the private sector, rather than the public, is responsible for products at the end of their life, giving companies an incentive to make their products last longer. The beginning of the cycle, not the end, might be when we can most effectively eliminate trash.

A Brief Chronicle of Garbage

1934: The Supreme Court bans cities from dumping garbage into the ocean.

1905: One of the first incinerator in the US begins powering the lights on New York's Williamsburg Bridge.

3000 B. C.: The earliest recorded landfill is built on the island of Crete.

1942: The War Production Board urges Americans to conserve resources with its "Get in the Scrap!" campaign.

1973: The nation's first curbside recycling program is introduced in Berkeley, California.

2002: Ireland becomes the first country to impose a tax on plastic shopping bags.

2150: The US recycles or composts 100 percent of its trash.

Notes:

① pneumatic tube 气动导管;输气管;气压管

② anaerobic digestion 厌氧消化;厌气消化

③ microbe 微生物,细菌

④ incinerator 焚化炉

⑤ plasma gasification 等离子体气化

⑥ pyrolysis 热分解

⑦ obsolescence 荒废;退化

Exercises

Ⅰ. Read each of the following statements carefully and decide whether it is true or false according to the text.

1. Many scientists, environmentalists and sanitation experts have presented creative ways to tackle the problem of trash. (　　)
2. Putting a price on trash doesn't work if people dump illicitly. (　　)
3. Some companies are looking for new uses for leftovers. (　　)
4. Plasma gasification and pyrolysis are expensive but efficient in turning trash into fuel. (　　)
5. Garbage can be used to build skyscrapers. (　　)
6. Ireland was the first country to impose a tax on plastic shopping bags. (　　)

Ⅱ. Answer the following questions according to the text.

1. How many ways are presented in this text to stop humans from filling the world with trash?
2. How do pneumatic tubes work to deal with trash?
3. What's the problem with garbage burning in the past?
4. What is plasma gasification?
5. What's the best way to end trash for good?

Ⅲ. Translate the following terms into their Chinese or English equivalents.

1. RFID
2. compost
3. facial-recognition software
4. recycling bin
5. incineration plants
6. 劳动密集型
7. 摩天大楼
8. 废物管理
9. 热分解
10. 不可回收材料

Ⅳ. Translate the paragraph into Chinese.

If rocket technologies improve, Staley says, we might one day blast trash into space and use the sun's heat to burn it. But given that our planet has limited resources, burning them after one use probably isn't the answer. Some environmentalists want to prevent companies from making nonrecyclable materials in the first place, and a few have suggested alternatives. A European research group designed a laptop made of recycled materials whose components can be reused.

Text B Surviving on Earth

No matter what we do to control our fossil-fuel use and carbon output, our climate has already been permanently charged for the next millennium. To prevent the planet from becoming uninhabitable, we'll have to take our control of the environment a step further and become geoengineers①, using technology to shape geological processes. Though "geoengineering" is the proper term here, I use the word "terraforming" because it refers to making other planets more comfortable for humans. As geoengineers, we aren't going to "heal" the earth or return it to a prehuman "state of nature." That would mean submitting ourselves to the vicissitudes of the planet's carbon cycles, which have already already caused several mass extinctions. What we need to do is actually quite unnatural: we must prevent the earth from going through its periodic transformation into a greenhouse that is inhospitable to humans and the food webs where we evolved. Put another way, we need to adapt the planet to suit humanity.

To make earth more human-friendly, our geoengineering projects will need to cool the planet down and remove carbon dioxide from the atmosphere. These projects fall into two categories. The first, called solar management, would reduce the sunlight that warms the planet. The second, called carbon-dioxide removal, does exactly what it sounds like.

Futuristic Jamais Cascio, author of *Hacking the Earth: Understanding the Consequences of Geoengineering*, predicts that we'll see an attempt to initiate a major geoengineering project in the next 10 years, and it will probably by solar management. "It's a faster effect and tends to be relatively cheap, and some estimates are in the billion-dollar range," he said, "It's cheap enough that a small country or a rich guy with a hero complex who wants to save the world could do it." Indeed, one solar-management project is already under way, albeit inadvertently②. Evidence suggests that sulfur-laced aerosol③ exhaust emitted by cargo ships on the ocean changes the structure of high clouds, making them more reflective and possibly cooling temperatures over the water. Some solar-management plans take note of this discovery and propose that we fill the oceans with ships spraying aerosols high into the air. But other strategies are more radical.

To find out how we'd shield our planet from sunlight, I visited the University of Oxford, winding my way through the city's maze of pale gold spires and stone alleys to find an enclave of would-be geoengineers. Few deliberate geoengineering projects have been tried to date, but the mandate of the future-focused Oxford Martin School is to tackle scientific problems that will become important over the next century. One of its researchers is Simon Driscoll, a young geophysicist who divides his time between studying historic volcanic eruptions and figuring out how geoengineers could duplicate the effects of a volcano in the earth's atmosphere without actually blowing anything up.

Driscoll told me about what volcanoes do to the atmosphere while cobbling together cups of tea in the cluttered atmospheric-physics department kitchen. Along with all the flaming lava④, they emit tiny airborne particles called aerosols, which are trapped by the earth's atmosphere. He cupped his hands into a half sphere over the steam erupting from our mugs of tea, pantomiming the layers of earth's atmosphere trapping aerosols. Soot⑤, sulfuric acid mixed with water, and other particles erupt from the volcano and shoot far above the breathable part of our atmosphere, but remain hanging somewhere above the clouds, scattering solar radiation back into space. With less sunlight hitting the earth's surface, the climate cools. This is exactly what happened after the famous eruption of Krakatoa in the late 19th century. The eruption was so enormous that it sent sulfur-laced particles high into the stratosphere⑥, a layer of atmosphere that sits between six and 29 miles above the planet, where they reflected enough sunlight to lower global temperatures by 2. 2 degrees Fahrenheit on average. The particles altered weather patterns for several years.

Driscoll drew a model of the upper atmosphere on a whiteboard. "Here's the troposphere⑦," he said, drawing an arc. Above that he drew another arc for the tropopause, which sits between the troposphere and the next arc, the stratosphere. Most planes fly roughly in the upper troposphere, occasionally entering the stratosphere. To cool the planet, Driscoll explained, we'd want to inject reflective particles into the stratosphere, because it's too high for rain to wash them out. These particles might remain floating in the stratosphere for up to two years, reflecting the light and preventing the sun from heating up the lower levels of the atmosphere, where we live. Driscoll's passion is in creating computer models of how the climate has responded to past eruptions. He then uses those models to predict the outcomes of geoengineering projects.

Harvard physicist and public-policy professor David Keith has suggested that we could engineer particles into tiny, thin discs with "self-levitating" properties that could help them remain in the stratosphere for over 20 years. "There's a lot of talk about 'particle X,' or the optimal particle," Driscoll said. "You want something that scatters light without absorbing it." He added that some scientists have suggested using soot, a common volcanic by-product, because it could be self-levitating. The problem is that data from previous volcanic eruptions show that soot absorbs low-wavelength light, which causes unexpected atmospheric effects. If past eruptions like Krakatoa are any indication—and they should be—massive soot injections would cool most of the planet, but changes in stratospheric winds would mean that the area over Eurasia's valuable farmlands would get hotter. So the unintended consequences could actually make food security much worse.

It's not clear how we'd accomplish the monumental task of injecting the particles, but getting particles into the atmosphere isn't the tough part. The real issues, for Driscoll and his colleagues, is the unintended consequences of doping our atmosphere with substances normally unleashed during horrific catastrophes. Rutgers atmospheric scientist Alan Robock has run a number of computer simulations of the sulfate-particle injection process and warns that it could destroy familiar weather patterns, erode the ozone layer, and hasten the process of ocean acidification⑧, a major cause of extinctions. "I think a lot about the doomsday things that might happen," Driscoll said. Unintended warming and acidification are two possibilities, but geoengineering could also "shut down monsoons⑨," he speculated.

Nonetheless, if the planet starts heating up rapidly, and droughts are causing mass death, it's very possible that we'll become desperate enough to try solar management. The planet would rapidly cool a few degrees and give crops a chance to thrive again. What would it be like to live through a geoengineering project like that? "People say we'll have white skies—blue skies will be a thing of the past," Cascio said. Plus, solar management is only "a tourniquet," he warned. The greater injury would still need treating. We might cut the heat, but we'd still be coping with elevated levels of carbon in our atmosphere, interacting with sunlight to raise temperatures. When the reflective particles precipitated out of the stratosphere, the planet would once again undergo rapid, intense heating. "You could make things significantly worse if you're not pulling carbon down at the same time," Cascio said.

That's why we need a way of removing carbon from the atmosphere while we're blocking the sun. One of the only geoengineering efforts ever tried was aimed at pulling carbon out of the atmosphere using one of the earth's most adaptable organisms: algae called diatoms⑩. In several experiments, geoengineers fertilized patches of the Southern Sea with powdered iron, creating a feast for local algae. This resulted in enormous algae blooms. The scientists' hope was that the single-celled organisms could pull carbon out of the air as part of their natural life cycle, sequestering the unwanted molecules in their bodies and releasing oxygen in its place. As the algae died, they would fall to the ocean bottom, taking the carbon with them. During many of the experiments, however, the diatoms released carbon back into the atmosphere when they died instead of transporting it into the deep ocean. While a few experiments have suggested that carbon-saturated algae⑪ can sink to the ocean floor under the right conditions, the jury on this option is still out.

Another possibility would be to enlist the acid of rocks. One of the most intriguing theories about how we'd manipulate the earth into pulling down carbon was dreamed up by Tim Kruger, who heads the Oxford Martin School's geoengineering efforts. I met with him across campus from Driscoll's office, in an enormous stone building once called the Indian Institute and devoted to training British civil servants for jobs in India. It was erected at the height of British imperialism, long before anyone imagined that burning coal might change the planet as profoundly as colonialism did.

Kruger is a slight, blond man who leans forward earnestly when he talks. "I've looked at heating limestone⑫ to generate lime that you could add to seawater," he explained in the same tone another person might use to describe a new recipe for cake. Of course, Kruger's cake is very dangerous—though it might just save the world. "When you add lime to seawater, it absorbs almost twice as much carbon dioxide as before," he continued. Once all that extra carbon was locked into the ocean, it would slowly cycle into the deep ocean, where it would remain safely sequestered. An additional benefit of Kruger's plan is that adding lime to the ocean could also counteract the ocean acidification we're seeing today. Given that geologists have ample evidence that previous mass extinctions were associated with ocean acidification, geoengineering an ocean with lower acid levels is obviously beneficial. "A caveat is that we don't know what the environmental side effects of this would be," Kruger said, echoing the refrain I'd already heard from Driscoll and others.

Another possible method of pulling carbon down with rocks is called "enhanced weathering." We can see a model for this in the Ordovician Period⑬, about 450 million years ago. During this period, intense weathering from wind and rain wore the Appalachian

Mountains down to a flat plain; runoff from the shrinking mountain took tons of carbon out of the air, raising oxygen levels and sending the planet from greenhouse to deadly icehouse. Cambridge physicist David MacKay recommends this form of geoengineering in his book *Sustainable Energy: Without the Hot Air*. "Here is an interesting idea: pulverize⑭ rocks that are capable of absorbing CO_2, and leave them in the open air," he writes. "This idea can be pitched as the acceleration of a natural geological process." MacKay imagines finding a mine full of magnesium silicate⑮, a white, frangible mineral often used in talcum powder⑯. We'd spread magnesium-silicate dust across a large area of landscape or perhaps over the ocean. Then the magnesium silicate would quickly absorb carbon dioxide, converting it to carbonates that would sink deep into the ocean as sediment.

However we do it, we need to begin to maintain the climate at a temperature that's ideal for human survival. Instead of allowing the planet's carbon cycle to control us, we would control it. We would adapt the planet to our needs by suing methods learned from the earth's history of extraordinary climate changes and geological transformations. We'll also need to adapt the climate to serve the creatures that share the world's ecosystems with us. If we want our species to be around for another million years, we have no choice. We must take control of the earth. We must do it in the most responsible and cautious way possible, but we cannot shy away from⑰ the task if we are to survive.

Notes:

① geoengineer 由 geoengineering 衍生而来，geoengineering 意为"地球工程"，是指采取人工反射阳光或者人工捕捉和存储温室气体的方式迅速逆转全球变暖趋势

② inadvertently 漫不经心地；疏忽地；非故意地

③ sulfur-laced aerosol 含硫化物的气溶胶

④ lava 熔岩

⑤ soot 烟尘

⑥ stratosphere 平流层

⑦ troposphere 对流层

⑧ ocean acidification 海洋酸化

⑨ monsoon 季风

⑩ diatom 硅藻

⑪ carbon-saturated algae 碳饱和藻

⑫ limestone 石灰石

⑬ Ordovician Period 奥陶纪，地质年代名称，是古生代的第二个纪（原始的脊椎动物出

现),开始于距今488个地质单位之前(4.8亿年前)至444个地质单位之前(4.4亿年前),延续了4 200万年。在奥陶纪与志留纪之间隔着一起大规模物种大灭绝——伽玛射线暴。在此次物种大灭绝中,60%的物种灭绝,主要灭绝的动物有:圆月形镰虫、彗星虫、原始生物

⑭ pulverize 将……弄碎;将……弄成粉末或尘埃;摧毁;粉碎

⑮ magnesium silicate 硅酸镁

⑯ talcum powder 滑石粉

⑰ shy away from 逃避;退缩

Exercises

Work in groups and discuss the following questions.

1. What can the geoengineering projects do to make earth more human-friendly?
2. What's a stratosphere?
3. Why can injecting reflective particles into the stratosphere cool the earth?
4. Which ways can be used to remove carbon from the atmosphere?
5. What's enhanced weathering?

科技英语翻译

科技文体涵盖的领域很广,当今世界科学技术日新月异,中外科技交流非常频繁,科技翻译在其中起着举足轻重的作用。本章就科技文体的翻译要点与技巧进行介绍。

第一节 科技英语翻译概论

在任何翻译活动开展之前,必须明确翻译的目的。

翻译的目的一般可分为三个方面:

1. 译者自身的目的 ;
2. 译文在译语文化中的交际目的;
3. 译者采用某种翻译方法、翻译策略或翻译形式所要达到的目的。

就翻译的本质而言,起主要作用的是译文在译语文化中所要达到的交际目的。

德国功能目的论(skopos theory)奠基人之一威米尔(H. J. Vermeer)对他的目的性原则(skopos rule)的陈述如下:

每种语篇的产生都有一定的目的,并服务于这一目的。目的性原则是这样表述的:无论笔译、口译、讲话或写作所生成的语篇/译文,都要能在对方的语境中对想要使用该语篇/译文的人确切地发挥它的功能。

在威米尔等人的目的论中,翻译的目的和功能被提升到至高无上的地位。翻译工作的具体目标往往跟翻译的发起人或委托人(initiator/commissioner)的特定目的联系在一起,也就是说译文要满足委托人的要求,“要能在对方的语境中”为受众或使用者所接受,“确切地发挥它的功能”。

翻译目的确定后,译者就可根据具体的翻译目的,确定翻译的策略,开展翻译活动。

在科技翻译中,多数的情况是为目的语读者提供一种自然的译文,即采取归化翻译法。归化翻译法旨在尽量减少译文中的异国情调,为目的语读者提供一种自然流畅的译文。美国著名翻译理论学家劳伦斯·韦努蒂认为,归化法源于这一著名翻译论说,“尽量不干扰读者”。

而与之相对应的异化法,则迁就外来文化的语言特点,吸纳外语表达方式,目的是使译文保持原文的文化风貌,让读者熟悉原语文化。

可见,在了解翻译的目的后,译者主体的本质表现在其能动性、受动性、为我性的特征中,而这些特征构成了翻译主体译者的主体性。译者具有翻译过程中的主体性地位,要充分发挥主观能动性,对文本信息的量与质的综合传递给予相应的控制和调节。

译者主体性即译者的主观能动性。译者的主观能动性与受动性相对。

译者能在多大程度上克服受动性,就能在多大程度上发挥主观能动性。译者要认识其主体作用,要认识翻译工作的创造性,在可能的再创造空间中自由驰骋;要增强翻译职业素养以充分发挥主观能动性。

一、译者的社会角色

功能目的论认为,翻译是一项有目的的(intentional)交际活动。在整个翻译过程中起主导作用的是译文在译语文化中所要达到的交际目的。具体地说,译者受委托人(包括社团及读者等)的委托,接受翻译任务,同时也就明确了翻译的具体目的,所以译者对翻译采取前瞻的态度(prospective attitude)。

钱锺书先生曾有以下体会:

一国文字和另一国文字之间必然有距离,译者的理解和文风跟原作品的内容和形式之间也不会没有距离,而且译者的体会和他自己的表达能力之间还时常有距离。从一种文字出发,积寸累尺地度越那许多距离,安稳到达另一种文字里,这是很艰辛的历程。(选自《林纾的翻译》)

二、译者的创造空间

1. 再造想象

再造想象是按照语言文字的描述或图样、符号、标记、模型等的示意,在头脑中再现出与之相应的事物形象的心理过程。

再造想象的根本特征是"再现"或"复制",它要与原对象保持最大限度的一致。

1.1 审美再造想象

审美再造想象不仅允许而且要求审美主体充分发挥能动作用,获得与创造者主观意图不相一致的审美结果。

文学翻译固然需要有更多的审美再造想象,科技翻译,特别是科普翻译,也经常需要张开想象的翅膀。

原文:(After resting on the ocean floor, split asunder and rusting, for nearly three-quarters of a century, a great ship seemed to come alive again.)... As they viewed videotapes and photographs of the sunken leviathan, millions of people around the world could sense her mass, her eerie quiet and the ruined splendor of a lost age.

译文:(一艘巨轮在撞击断裂后沉入海底70多年,锈蚀不堪,似乎又要获得新生。)……当世界上千百万人观看这沉没巨轮的录像带和照片时,他们都会感受到它那庞大的躯体,察觉到它周围可怕的沉寂,领略到它已经剥落的昔日的辉煌。

以上译文就是根据巨轮沉没的年代、巨轮的大小等信息,加上基于科学的审美想象出的译文,既符合事实又赋予读者充分的想象空间。

1.2　科学再造想象

科学再造想象要求客观、明确、精密，主体能动性的发挥要沿着创造者的思路探行，必须一丝不苟，对工程图纸等非语言表达的再造想象是这样，对文字描述也是这样。

原文：We shall design a barrel through which projectiles are to be fired comprised of a metal barrel shaped substrate upon at least the inner surface of which is coasted a relatively thin layer of a dense, hard wear and corrosion resistant vapor-deposited material having a Moh's scale hardness of about 8 to 9.

译文：我们将设计一种发射子弹的枪筒，它有一个圆筒形金属衬套，至少要在衬套内壁镀上一层致密耐磨且防蚀的气相淀积薄膜，其莫氏硬度约为8—9。

原文客观细致地描述了枪筒内壁衬套的加工，译文很科学地将衬套内壁的材质、材质的特性以及加工工艺作了全面的描述，客观而缜密。

2. 设立新名

随着科学技术的发展，不断地有新的术语产生，而术语的翻译需要遵循一定的原则，具体来说有以下几点：单义性、科学性、通俗性、系统性、国际性等原则。

以纳米相关术语为例。纳米为一长度单位，是 nanometer 的译名。纳米科技是迅速发展起来的新兴科技，随之产生了一系列的新术语。英语中和纳米相关的术语都以 nano- 开始，设立汉语中相应术语时也用"纳米" 开头，于是有了纳米技术、纳米电子学、纳米精度、纳米级、纳米材料、纳米金属等一系列各自意义明确、通俗易懂以及成体系的术语。

归纳翻译外国科技术语的方法，约有10种，即意译、音译、半音半意译、音译兼义译、形译、借用日语汉字、按外文字母译、造新字、简称、直接采用外文缩写词等，其中大多要再创造。

3. 转换语言形式

翻译中的再创造是指译者克服字面意义的羁绊，译出其指称意义；或者克服语言形式的羁绊，译出其言内意义；或者按语境的需要译出语用意义。

原文：His mind swept easily over the whole world including many lands and epochs.

译文：他总揽全球，古今中外，了如指掌。

原文中 mind 本义为"头脑""心思"，根据此句中的内涵把它译为心智、智慧更贴切。

三、译者素养

在当代翻译领域里，译者的职业素养最终表现为译者的职业能力或简称译者能力（translator competence）。

在当今时代，译者能力可概括为译者中外文功底、相关专业知识以及信息与通信技术的掌握。

1. 外语素养

外语不精,不可能正确理解以外语为原文的思想内容,更谈不上把握作者的语言风格、神韵色彩和思想脉络。许多错译即源于此。

原文:Each patient is an individual with different needs, depending on his or her special illness or condition.

原译:根据他或她各自的不同疾病或情况,每个病人都是有不同需要的个体。

此译文看似忠实于原文,但是不符合汉语表达习惯。于是遵照原文,进行改译。

改译:每个患者的病情或状况不同,需求亦不同。

2. 汉语素养

加强汉语素养,目的主要在于不断提高译者的汉语语言能力,提高语言对比的起点,提高汉语的表达能力。

原文:Air will completely fill any container which it may be placed in.

原译:空气将完全填满可以在其内放入空气的任何容器。

读此译文,丝毫体会不到要表达的意思,所以修改译文如下。

改译:空气会充满容纳它的任何容器。

此外,汉语有许多四字词组,适当应用,可使译文简洁精练,如:

Nothing, whether a weighty matter or a small detail, was over-looked.

译文:事无巨细,无一遗漏。

3. 专业知识

这里所说的专业知识,一方面是指对翻译的理性认识,以及一定的翻译技能和熟巧程度,另外科技翻译涉及的专业和领域非常广泛,所以译者还应具备翻译所对应专业和领域的相关知识。

原文:By the age of 19 Gauss had discovered for himself and proved a remarkable theorem in *number theory* known as the law of quadratic reciprocity.

原译:高斯19岁时已经独立地发现并证明了数理定律——二次互反律。

而如果对数学有一定的知识储备和了解,就会知道原文中的 *number theory* 是指数论,而非数理定律。

改译:高斯19岁时已经独立地发现并证明了数论中的一个卓越的定律,名为二次互反律。

原文:Temperature required for annealing is a function of two factors, (1) the nature of material, (2) the amount of *work* that has been done prior to annealing.

译文:退火所需的温度随两个因素而变:(1) 材料特性;(2) 退火前的加工量。

原文中的 work 的确切含义也是必须具备热加工与热处理相关知识才能准确地把握，从而避免错译。

4. 信息与通信技术

当代，信息量越来越大，传递速度越来越快。与之相适应的翻译量与日俱增，翻译速度成倍增长，翻译的无纸化程度急速提高。译者要利用网络与委托人沟通，通过网络试译、了解翻译要求、接收原文、传递译文。译者要利用委托人的数据库和自己的数据库在计算机上从事翻译工作。

信息与通信技术的掌握使译者能力进一步提升。目前，在经贸、法律、科技等方面，社会对优秀翻译工作者的需求量很大，但是形成极大反差的是，不少富有经验的退休老翻译却译笔高搁，无所事事。其原因主要不是精力不济，而是不懂计算机，缺乏计算机应用能力。

第二节 译品的类型

有观点认为,按照不同的分类标准,翻译可以分为多种类型。按照文本类型来分,翻译可分为科技翻译、文学翻译、商务翻译、经贸翻译等;按照处理方法来分,翻译可以分为全译、编译、摘译、节译等。这种按照处理方法的分类标准,也可以说是对原作内容与形式的保留程度的分类标准。那么,根据这种标准,翻译可以分为力求保全的“全译”和有所改变的“变译”。变译的实质是译者根据特定条件下特定读者的需求采用增、减、编、述、缩、并、改等变通手段摄取原作有关内容的翻译活动。

在我国翻译史上,翻译运作的主流应该说对原文信息总体上不增不减或力求保全,但是也存在对原作有增有减、有评有写,甚至完全改头换面的翻译作品。例如 19 世纪末严复、林纾等翻译家翻译的作品,因为各种原因几乎都有增删。随着信息时代的到来以及接受美学的盛行,人们越来越重视如何根据自己的需求对数量庞大的信息进行选择,因而对原作信息有所取舍、有所改造以适应市场需求和读者要求的翻译逐渐占据主导地位,成为跨文化信息传播的主要方式。

一、全文翻译

全文翻译(complete translation),顾名思义,就是指译者对原文或原著完整地翻译,不加任何删节。全文翻译是针对翻译操作对象的完整程度而言的。纵观翻译史,全文翻译一直都是翻译实践的主流。

即使在信息爆炸时代,全文翻译仍然是翻译实践中不可或缺的一个组成部分,尤其是文学作品翻译。

1. 操作原则

全文翻译必须强调再现原作内容和形式的完整性和整体性,而这一点也同时对译者创造性的发挥带来一定的限制。既然全文翻译的变化在于“微观调整”,那么译者的自由度也只存在于微观层次,而非在宏观结构上的很大改动。

2. 适用范围

全文翻译适用范围最广,任何语体、任何文体的作品都可以用全文翻译的方法用另一种语言表达出来。政治外交场合、法律文书、严肃文学作品等尤其需要完整无缺地再现原作的思想内容。在某种程度上,全文翻译比变译难度低,因为全文翻译不需要译者对原著作结构上的调整或者编辑,从而降低了在信息重组调度上的难度。但这并不意味着全文翻译是一件轻而易举的事情。译者必须把握原文的内容与精神,全力以赴以达到在译文中保全完整的原作信息。

3. 对译者要求

全文翻译不要求译者对原作内容作任何改变,所以不需要译者作任何概括、编辑、改写等。它对译者提出的要求是普遍性的翻译要求,即译者需要具备的翻译能力,如双语理解和表达能力、翻译转换的技巧、相关领域的背景知识等。同时也要求译者不能对原作进行改动,限制译者主体性的发挥。

4. 例子分析

原文:

Building Bridges to the Future—with Soya

[1] Scientists in the United States say they have found a way to make tanks, tractors, cars and even bridges out of soya beans.

[2] A University of Delaware team called Acres (Affordable Composites from Renewable Sources) has filed a patent for a process that could end with cheaper, lighter and greener materials. They will replace the existing composites—plates, planks or beams of lightweight but costly petroleum resins, reinforced with glass fibers—to make material as stiff as steel but lighter.

[3] The first industrial partner is the US tractor firm Deere and Co., which predicts a $ 50 million market for farm equipment made out of soya.

[4] Soya is one of the world's great crops: the US alone grows more than 60 million tons a year. Soy products get into mayonnaise and medicines, anti-corrosion varnishes, fungicides and shampoos. Soy is used as a milk substitute, as cat food and as a cosmetic.

[5] Soya protein is being tested in Illinois as an oestrogen substitute and North Carolina scientists think it could contain a cancer prophylactic. Soya oil costs half as much as the polyester, epoxy and vinyl ester resins used in modern composite materials to make everything from ship hulls to aircraft parts.

全文翻译:

用大豆建造未来的桥梁

[1] 美国科学家声称,他们已找到了利用大豆生产坦克、拖拉机、轿车甚至是桥梁的方法。

[2] 特拉华大学的一个称为Acres(从再生资源中提取可利用合成物)的研究小组已申请了一项工艺的专利,将采用更加便宜、轻便和天然的材料来代替现有的复合物,如金属、木质板材或重量轻成本高的束状石油树脂(其中加入了玻璃纤维以提高强度),可使材料像钢

一样硬而重量则轻得多。

[3] 首家工业合作伙伴是美国 Deere 拖拉机公司，他们预测采用大豆为原材料生产农用设备的市场将达 5 000 万美元。

[4] 大豆是世界上产量最大的农作物之一：仅美国全年产量就超过 6 000 万吨。大豆制品涉及蛋黄酱和药品，防腐清漆，杀菌剂及洗发香波。大豆可用作牛奶的替代物、猫食及化妆品。

[5] 试验证明大豆蛋白质可代替雌性激素，其内部含有防癌物质。豆油成本只及聚酯、环氧树脂及乙烯基酯树脂的一半，它们用在现代复合材料中以生产从船体到飞机零件等各种东西。

分析：

1. 之所以采用变体，是为了吸引读者，"豆腐渣工程"在中国已为人熟知，采用悬念可以显示新技术的独特之处。另外，题目虽说造桥，实际上不限于此，而是介绍大豆可用作许多工业领域的生产材料。

2. 具体的研究单位大众读者不太感兴趣，那是某些科技工作者的兴趣，只需说一说其生产原理即可。

3. 段 3 属于未来工作，应调后。

4. 段 4 与段 5 可合并，只突出大豆现有的工业作用即可，与食品相关的可舍去。

于是，就有了如下的翻译变体：

用大豆造桥

[1] 美国科学家声称，大豆还可用来制造坦克、拖拉机、轿车，甚至是桥梁。

[2] 大豆可用来制蛋黄酱、药品、防腐清漆、杀菌剂、洗发香波等。大豆蛋白质可代替雌性激素，含有防癌物质。豆油成本只及聚酯、环氧树脂及乙烯基酯树脂的一半，可以生产从船体到飞机零件等各种东西。

[3] 大豆用作桥梁桁架及干草带编织机镶板的生产材料尚属首次。材料科学家在豆油中加入化学物质和玻璃纤维即可韧化豆油，使之坚硬如刚且重量较轻。

[4] 1997 年 11 月，一种新型大豆复合材料研制成功：这是一个用大豆制成的玻璃钢编织机零件，面积为 2.5 m×1 m，重 11 公斤，仅为同类金属零件重量的四分之一。

[5] 产研结合的首家工业伙伴是美国 Deere 拖拉机公司，他们预测采用大豆为原材料生产农用设备的市场将达 5 000 万美元。

二、节译

节译（selected translation）指一部著作或一篇文章局部地被删节了的翻译。另外，当原著中出现了少量不宜翻译出来的内容的时候，译者可以适当地删去不译，仅在译文的前言或后记里注明“略有删节”。应该说，节译和“略有删节”是有区别的。节译将一本书或一篇文章的全部或部分节缩译出，其翻译要求与全文翻译是完全一样的，差别只在于因为某些原因，全文不需要全部译出而已。节译中删节的内容应比“略有删节”的翻译更多，不是因为不宜翻译，而是因为根据信息重要性而作出的取舍。

1. 操作原则

节译也要遵循一定的原则。例如针对性原则，同样也是对翻译内容的取舍问题。取舍必须有针对性，针对译文的目标读者，尽量满足读者的期望和需求。此外，还需要遵循重要性原则。节译所选择的内容必须有所侧重，不能随心所欲。除了满足读者的需求外，还必须体现普遍的价值意义。所以，译者必须花费一定的时间和精力衡量和选取翻译的内容。

2. 适用范围

根据译者翻译的目的和节译的意义整体性特点，节译可适用于多种文体的翻译，例如文学作品的节选翻译、社科作品的选段翻译等。需要注意的是，节译的内容必须具有相当的整体性，能够独立成为思想意义相对完整的语篇，不能是支离破碎、含混不清的任意选译。另外，还需要在适当的地方注明所节选的具体出处，以便有兴趣的读者进一步阅读查找。

3. 对译者要求

节译者除了要具备基本翻译能力外，还要求译者具备对原作各部分内容的分析判断能力，能够对原作整体思想进行准确无误的把握。

三、改译

有翻译学者认为翻译就是改写(rewriting)。此处的改译并不是用来定义翻译的改写，而是译者在理解原作的基础上根据自己的理解重新书写的翻译方法，使原作在内容、形式，甚至风格上发生明显改变。这种方法赋予译者相当的自由，可以说是译者的重新创作。改译带有明显的译者特点。改译也可以是为了纠正原作的错误，或者直接纠正，或者按照原文直译，再加上注释说明错误之处。改译可以帮助读者理解原文，消除因为语言文化差异而引起的沟通问题。

1. 操作原则

改译也需要遵循一定的原则。首先，改译可以使原作在内容上、形式上，甚至风格上发生明显改变，而改变必须以原作为基础，违反了这条原则，改译是否为翻译都会成问题。

其次，译者在改译过程中享有充分的主体性，可以对原文进行一定的处理，但是必须以保证译文的整体性为前提。改写并不是随意的改动，不能把原文打得支离破碎。

2. 适用范围

改译可用于文学作品的译介，也可用于应用文体翻译。对于可以调整改变表达形式而不会对原作造成严重损害的语篇内容，改译是迎合读者喜好、赢取市场的好方法。需要注意的是，尽管改译赋予译者相当的自由度，译者始终要受到原文的约束。

3. 对译者要求

改译并不是轻而易举的事情，译者必须充分把握原作的精神，需要具备对原文理解与表达的能力、文学创作能力、对作品内容重组的能力等。

4. 例子分析

严复翻译赫胥黎《天演论》的开场白，是常引争议的例子。

原文：It may be safely assumed that, two thousand years ago, before Caesar set foot in southern Britain, the whole country-side visible from the windows of the room in which I write, was in what is called "there state of Nature". Except, it may be, by raising a few sepulchral mounds, such as those which still, here and there, break the flowing contours of the downs, man's hands had made no mark upon it; and the thin veil of vegetation which overspread the broad-backed heights and the shelving sides of the coombs was unaffected by his industry.

译文：赫胥黎独处一室之中，在英伦之南，背山而面野，槛外诸境，历历如在几下。乃悬想两千年前，当罗马大将恺彻未到时，此间有何景物。计惟有天造草味，人功未施，其借征人境者，不过几处荒坟，散见坡陀起伏间，而灌木丛林，蒙茸山麓，未经删治如今日者，则无疑也。

比读一下，不难看出严复解构了原作，又照汉语习见的方式重新组合，改复合句为若干短句，并列关系替代了主从关系。原文第一人称改为第三人称。原文以事实开头，译文以人开头。改译的原因，大概是为了读起来像中国古代的说部和史书，史书的开头多半是：太史公曰，臣光曰之类。这一开始就动摇了原作的风格，严译比原文更戏剧化。王佐良先生说原文首句是板着面孔开始的，而严译第一句就把读者带到了一个富有戏剧性的场合，而且本段略后处，严复将原文 unceasing struggle for existence 译作"战事炽然。强者后亡。弱者先绝。偏有留遗"。加了许多字，读起来简直像战况公报。科学理论著作翻译的戏剧化的动因，王佐良认为可能是严复想增强译作强烈的历史意识，利用种种风格手段以增强读者的历史感。

四、编译

从本质上说，编译（adapted translation）就是把一种语言所写的一篇文章或一本著作的内容，或者若干篇文章、若干部著作中的相关内容，用另一种语言忠实而又相对完整地概述出来。编译可以用简洁凝练的语言和较小篇幅在较短时间内译介大量的内容，具有速度快、效率高、实用性强等特点。编译具有相对的自由性，但是并不意味着编译可以随心所欲，篡改原文，更不可以看得懂的就译，看不懂的就删掉。编译必须紧扣原作的内容，取其精华，以凝练的语言集中表达所翻译作品的内容和中心思想。编译有广义和狭义之分。广义的编译包含了译者个人的观点，而狭义的编译不能增加译者自己的观点，译者的工作只是编译原作的精华部分加以重组、翻译。

1. 操作原则

编译在摘译的操作原则上还可增加概括性原则、调整性原则、层次性原则以及合并性原则等。

概括性原则。即编译者必须对原文内容进行高度概括,要脱离原文的语言形式,把原文的精神概括提取出来加以翻译。

调整性原则。仅仅把原文的内容高度概括还不足以得到好的编译。概括了内容后,还需要对内容作出调整。由于不同国家的语言文化、风俗习惯多种多样,在翻译中对原文材料作出适当的调整以适应译入语读者的需要是编译的一个重要原则。

层次性原则。编译是把若干原著的相关内容经过概括调整后进行翻译的活动。各相关的内容必须以逻辑明确、层次分明的形式出现。编译的层次性原则就体现在译文表达的条理上。

合并性原则。编译既然把各种相关的内容经过整合后进行翻译,那么,译文必须以统一整体的形式出现。换句话说,原文的相关内容必须在翻译过程中进行融会贯通,各部分合成不可分割的整体。否则,编译的译文只会是原文各种内容的机械罗列,逻辑混乱的编译绝不是好的译文。

2. 适用范围

由于具有速度快、效率高、实用性强的特点,编译广泛应用于各种领域语篇的处理,例如科技情报报道、科研综述、经济社会报道、新学说新思想的译介等。在信息爆炸的时代,读者要求在有限的时间获取大量的有用的信息,编译这种特殊的翻译处理方法将会更加广泛应用于各个领域。翻译的时候需要注意向读者提供编译的文章来源、编译的倾向性、编译的取舍原则等。

3. 对译者要求

编译集翻译与编辑于一身,不仅要求译者有较强的外语阅读能力,做到吃透原作的内容,掌握它的要领;还要求能用另一种语言忠实而又通顺地把它再现出来。所以,编译要求译者具有深厚的语言功底,尤其是汉语表达能力。其次,编译要求译者具有丰富的相关专业知识。在编译过程中,译者不仅仅要理解原作的字面含义,还要理解原作所涉及的基本原理和行业知识。最后,编译还要求译者具备一定的鉴赏能力、分析判断能力和一定的编辑能力。编译既有编辑又有翻译,译者除了要有翻译能力外,还要掌握编辑的技巧。编译者必须能够从纷繁的材料中编辑出符合读者期望的内容,把有价值的部分翻译给读者。

4. 例子分析

原文:

Hormones Could Curb Broken Hips in Women

By Oliver Gillie

Medical Editor

[1] WOMEN who have passed through the menopause should consider taking extra female hormones to prevent bone fractures, the Royal College of Physicians said yesterday.

[2] Its advice follows a two-fold increase in hip fractures in old people over the past 20 years. The increase is probably the result of changes in lifestyle, particularly a decline in the amount of exercise people take.

[3] More than 46,000 people a year suffer hip fractures in England and Wales and about a quarter die as a result. Many more never regain full mobility. The hospital costs are about £ 160m a year, the college says in a new report.

[4] Women are most vulnerable after the age of 50 because calcium is most from bones more rapidly when female hormones decline after the menopause. Hip fractures are most common in older women; almost 60 percent occur in women over 75.

[5] Middle-aged women vary in the amount of calcium lost from their bones; one third suffer heavy losses, making them much more vulnerable to fractures. These women may be detected by screening methods which involve shining a light beam through the wrist, but this method is not widely available.

[6] Dr. John Kanis, reader in human metabolism at Sheffield University, said: "Female hormones help to prevent loss of calcium from bones. After the menopause some female hormone is still produced in body fat but thin women produce very little female hormone and so they should have hormone replacement therapy [extra female hormones].

[7] "Women who have their ovaries removed surgically and women who have a premature menopause [while they are in their early 40s or younger] should have hormone replacement therapy because they start losing calcium from their bones earlier.

[8] "Heavy smokers and heavy drinkers should also have hormone replacement therapy because they produce less female hormone naturally. "

[9] However, medical opinion differs on the therapy. Sir Raymond Hoffenberg, president of the college, said: "It is advisable for all women over 50 to have hormone replacement therapy, except for certain women who should not take it because here is an increased risk of breast cancer

if close relatives have had the disease. ”

[10] However, loss of calcium can also be prevented by taking exercise and eating foods rich in calcium (best sources: milk and milk products, canned fish eaten with bones, dark-green leafy vegetables).

[11] Fractured Neck of Femur: Prevention and Management. Royal College of Physicians, 11 St Andrews Place, London, NW14LE. (£ 5)

原编译稿:

激素可防止女性髋部骨折

[1] 不久前,英国皇家医学院(The Royal College of Physicians)的专家们公布,绝经后的妇女可采用补充雌性激素疗法来预防髋部(组成骨盆的大骨,左右各一,形态不规则,统称髋骨)骨折。

[2] 通过长期临床观察和大量研究,这些医学家们发现:雌性激素有助于防止骨骼中钙质的损耗;而50岁左右的妇女绝经期后雌性激素的分泌越来越少,尤其是体质瘦弱的妇女分泌量几乎是微乎其微。这就造成了骨骼内钙质的迅速损耗,导致髋部骨折屡屡发生。

[3] 据专家们测算,在过去的20年中,由于生活方式的改变和运动量的减少,英国老人中髋部骨折患者几乎增加了两倍。现在,在英格兰和威尔士两地,每年髋部骨折患者多达46 000余人,其中1/4因此而丧生,而绝大多数是绝经后妇女,75岁以上的老妇高达总发病率的60%左右。

[4] 最近,谢菲尔德大学人体代谢作用学高级讲师约翰·凯尼斯博士(Dr. John Kanis at Sheffield University)一再告诫那些做过卵巢摘除术以及绝经期提前(绝经期在40岁左右)的妇女和嗜烟酒成性的妇女要及早进行雌性激素补充疗法,以弥补体内雌性激素过度不足,防止骨骼中钙质的过早损耗而造成髋部骨折的发生。但是,英国皇家医学院院长雷蒙德·霍芬勃格爵士(Sir Raymond Hoffenberg)不同意近亲中有患乳腺癌的妇女接受这种新型疗法,以免诱发乳腺癌的危险。这些医学专家们一直认为,中老年妇女适当增加运动量,多食用牛奶、奶制品、深绿色叶蔬菜以及带骨头吃的罐头鱼等富含钙的食物,来补充骨骼中的钙质含量,也可起到防止骨折的作用。

[5] 据介绍,中年妇女骨骼内钙质的含量因人而异,其中1/3的人钙质损耗现象严重,极易发生骨折。目前,英国发明了一种新型查钙法(具体原理不详,属专利之列)——让一种特殊的光束通过腕部就可测定出骨骼内的钙质含量,以便及时对症下药进行治疗。但此测量法尚未普及。

见报编译稿：

激素可防妇女髋部骨折

［1］不久前，英国皇家医学院的专家们公布，绝经妇女可采用补充雌性激素疗法来预防髋部（组成骨盆的大骨）骨折。

［2］通过长期临床观察和大量研究，专家们发现，雌性激素有助于防止骨骼中钙质的损耗；而50岁左右的妇女雌性激素的分泌又越来越少，造成了骨骼内钙质迅速损耗，导致髋部骨折。在过去的20年中，由于生活方式的改变和运动量的减少，英国老人中髋部骨折患者几乎增加了两倍。在英格兰和威尔士两地，每年髋部骨折患者多达4万余人，而绝大多数是绝经后妇女。

［3］最近，谢菲尔德大学人体代谢专业高级讲师约翰·凯尼斯博士一再告诫那些做过卵巢摘除术、绝经期提前和绝经期后体瘦的以及烟酒成性的妇女要及早进行雌性激素补充疗法，防止髋部骨折。但是，英国皇家医学院院长雷蒙德·霍芬勃格爵士认为，近亲中有患乳腺癌的妇女不宜接受这种新型疗法，以免诱发乳腺癌的危险。为避免髋部骨折，医学专家们一致认为，中老年妇女适当增加运动量，多食用牛奶、奶制品、深绿色叶蔬菜以及带骨头吃的罐头鱼等含钙食物，来补充骨骼中的钙质含量。

分析：

比读原文与两篇编译稿，总的感觉是：不仅原文内容有删减，而且内容之间有融合，文章结构有调整，逻辑顺序更显豁。

在叙述逻辑上，编译者以时间为序，按事物发展的时间顺序报道国外医学科研成就。因为原稿发稿日期是1989年1月9日，等到国内报道出来已经是5月。为保证报道的科学性和严肃性，编译者在文章的时间脉络上分别加上了“不久前”（两篇编译的段［1］）和“最近”（原编译的段［4］和见报编译的段［3］）。

在文章结构上，出现了大调整。原作共11段，经编译者（此处包括原编译者和报社的编校者）的加工处理，见报编译稿只存3段。

内容上的融合，形式结构的变化，根本原因是编译者对所要表达的内容进行了调整。内容的表达决定形式的再现。请比较：

原编译的内容结构：总的报道→女性髋部骨折的原因、现象（含治疗方法）→治疗方法。

见报编译的内容结构：总的报道→女性髋部骨折现象及原因→治疗方法。

后者在内容上较之于前者显得紧凑，逻辑性强。

五、摘译

摘译（abstract translation）指译者根据实际需要摘取原文的中心内容或个别章节或段落进行翻译，其内容一般是原作的核心部分或者是译文读者感兴趣的部分内容。从程序上讲，摘译应是先摘，后取，再译，应该是译者根据特定的服务对象的需要，摘取原文的特定部分进行翻译的一种单向的跨文化传播活动。摘译的部分必须忠实于原文，不可加进译者的思想内容。摘译可以从各个层次上摘取原作内容，可分为删章译、删段译、删句译、删词译 4 种。严复译作中有很多都是摘译，采用上述 4 种摘译方法译介西方思想。

1. 操作原则

摘译在操作上可遵循以下原则：

整体性原则。部分是相对于整体而言的，只有了解了整体的结构，才能进行部分摘取。所以摘译者必须对原作有整体的把握，站在全篇的高度进行摘取，取舍有度。

针对性原则。摘译的目的无非是让本国读者了解域外风俗民情，所以摘译必须针对读者的兴趣和需要。

简要性原则。摘译者必须在理解原文主旨的前提下对摘取内容进行修整处理，使译文主体鲜明，文字简练，中心突出。

重要性原则。任何文章的内容都有主次之分，在摘译的时候必须选择重要的内容进行翻译，不能本末倒置，主次不分。

选择性原则。摘译的对象可能涉及多方面的内容，如果不能做到面面俱到，则要注意选择，主要是按照译者或者读者的兴趣优先选取。

客观性原则。摘译不能加入译者的个人观点，不能采取批判性的态度对原作进行评论。为了句段之间衔接自然，译者可以适当增加必要的关联词或句子，前提是对原作的客观处理。

2. 适用范围

摘译最适用于科技作品的翻译，是译介国外最新科技信息的一个有效方法。摘译者可以根据国内读者的需求摘取国外最新科技文献资料加以翻译，以促进国内科技发展，加快与国际发展接轨。摘译也可用于新闻报道等应用文体翻译，但是不太适用于文学作品的翻译，这是由文学作品固有的特点决定的。文学作品内容与形式不可分离，摘取部分并不能让读者欣赏到文学的美学效果。

3. 对译者要求

摘译作为变译的一种，译者必须注意对出现变化的地方做好标记，以方便读者查找。例如注明摘译文相对应的原文版面位置、增加连接纽带、用括弧加以区分、对所删部分打上标记等。

摘译者必须对读者的期望以及社会的期望有充分的了解，以此为出发点对翻译过程中取什么舍什么进行判断。译者除了具备基本的翻译能力外，还需具备相关领域的广博知识和对原作的分析判断能力。

4. 例子分析

原文：

Amazed Scientists Discover Space is Full of Vast Oceans of Water

[1] European scientists using an ultracold orbiting telescope have discovered unimaginable volumes of water in the space between the stars. The discovery raises new questions about life elsewhere in the universe—and provides new answers about why life was possible on Earth.

[2] The scientists were astounded to find water vapor in the freezing atmospheres of Jupiter, Uranus, Neptune and Saturn. They also revealed yesterday that they have detected water in the atmosphere of Saturn's mysterious moon Titan—which is to be visited by a joint US-European probe in 2004. They have even identified a cloud of water, less than a light year across, in the constellation Orion.

[3] Hydrogen is the raw material of the universe. Oxygen is made in huge quantities by the stars. So water should be no surprise. There have been fleeting glimpses of water vapor on the Sun's surface, and in March the Americans confirmed the presence of water on the Moon. But the discovery of vast oceans of water vapor—the mass of which in the Milky Way galaxy alone would equal that of tens of thousands of suns—by the European Space Agency's (ESA) infrared space observatory ISO astonished astronomers.

[4] "I think our imaginations failed to describe the variety of places, the variety of circumstances, and the variety of conditions under which we found it," said Paul Murdin, of the British National Space Center, yesterday.

[5] Using instruments called spectroscopes, scientists have identified an astonishing range of organic molecules in space, including hydrogen cyanide, alcohol and formaldehyde. But water is the first requirement for life. The £600 million ISO experiment confirms once again that there could be life on planets in other solar systems. The next step will be to look for tell-tale signs of life itself: concentration of oxygen or ozone in distant planetary atmospheres.

[6] The ISO discovery also confirms a growing belief that water on Earth may have been delivered by comets, icy scraps of dust left over from the birth of stars.

[7] The experiment opens a new window on the universe. More than 1,000 astronomers from Europe, Japan and the US have used the telescope, launched in 1995, for more than 26,000 observations. The data returned will keep researchers busy for years.

[8] "All the missions we undertake in ESA have to be unique," said Roger Bonnet, head of science at the space agency. "In principle, the Americans have done everything, so what remains

to be done is impossible. All our missions are Mission Impossible, but ISO is more impossible than any other."

[9] ISO should have survived for only 18 months but is still operating after 29. The telescope posed a huge technical challenge: to detect infrared rays, it had to be cooler than any object it looked at—and the average temperature of deep space is −270 ℃. So ISO had to be cooled with liquid helium which kept the instrument's temperature just above absolute zero (−273 ℃).

[10] The satellite has peered into clouds of space dust to watch the formation of solar systems, and deep into the heart of ultraluminous galaxies. It has also witnessed two galaxies colliding and, in one of them, the signature of a black hole sucking a blaze of stardust into itself. Besides mapping the edge of the known universe, it has also seen a star being born inside the Horsehead Nebula.

摘译译文：

科学家发现太空存在大量的水

[1] 欧洲科学家利用超冷轨道望远镜发现星际空间存在大量水。该发现向人类提出了宇宙其他地方是否存在生命的新问题，也为地球上存在生命提供了新的解释。

[2] 科学家惊奇地发现在木星、天王星、海王星和土星的低温大气中存在水蒸气。他们还透露已探测到土星的一颗神秘的卫星"泰坦"的大气中存在水，甚至确定猎户星座上存在云状水层。

[3] 氢是宇宙的基本元素，恒星又大量地产生氧元素，所以水的生成不足为奇。在太阳的表面观察到稍纵即逝的水蒸气。3 月份，美国人证实了月球上水的存在，而欧洲空间研究所(ESA)红外线太空天文台(ISO)发现大量水蒸气存在于银河系中，其规模相当于成千上万个太阳表面的水蒸气，这更令天文学家们为之一振。

[4] 利用光谱仪，科学家已发现太空中存在大量的有机分子，包括氰化氢、酒精和甲醛分子。耗资 6 亿英镑的 ISO 实验再次证实，在太阳系的其他星体上可能存在生命。下一步就是寻找生命存在的标志：遥远行星的大气中是否存有氧气或臭氧。

[5] ISO 的发现使人更加相信：地球上的水来自彗星、恒星形成时留下的宇宙尘埃冰屑。

分析：

很明显，原作10 个自然段，摘译后被删除了 5 个自然段：段[4]和段[7～10]。首先我们认为要介绍给国内读者的是"科学家发现太空存在大量的水"这一事实，以之为准绳，对原作进行删减，因此原段[2]中"which is to be visited by a joint US-European probe in 2004"一句省去，使译文专注现在的事实，不说将来的事。原段[2]第二句中的 yesterday 也可省略，这一明确时间概念在此显得并不重要，句意"他们还透露已探测到土星的一颗神秘的卫星'泰坦'的大气中存在水"足以反映这一信息,由于传输信息的翻译和出版时空障碍,对译文读者而言,"昨天"已成久远。

原段[4]是 the British National Space Center 的 Paul Murdin 说的话,可省。原段[7~10]写的是红外线太空天文台实验研究的意义、过程、时间、规模、技术要求和所取得的进展。这些过程、方式手段等均是发现太空存在水的背景知识,与我们要译介的事实结果相去甚远,我们感兴趣的只是科技信息,为大众读者提供快速传输科技成就发展。

译段[3]中"欧洲空间研究所(ESA)红外线太空天文台(ISO)"中的缩略语还可考虑保留与否,若是给大众看的综合性报刊,可省;若是给专业读者看的专业情报刊物,可保留。同理,原段[6]中的 ISO 也要还原为全称"红外线太空天文台"。

六、综译

综译(comprehensive translation)可以说是对一篇或多篇原作进行综述后再进行翻译的一种翻译方法。综译可能带上译者的评论讲解或者译者随意增删的内容。总的来说,翻译的种种变体,旨在提供充分利用国外各种信息的最有效、最快捷的引进方法。

1. 操作原则

综译的基本原则包括综合性和叙述性、真实性和准确性、充分性和必要性、新颖性和典型性、详略性和统一性、综译过程的研究性。

综合性和叙述性。所谓综合性,即纵向的综合和横向的综合,也可以说既从共时性又从历时性对某一领域进行综合分析研究。叙述性体现在对大量实际材料的客观性叙述。

真实性和准确性。综译所选择的材料必须真实可靠,并尽量选用第一手材料,以确保内容的准确性。此外,还必须明确区分何为客观叙述,何为主观评述。

充分性和必要性。面对浩如烟海的信息材料,选择时不可能做到面面俱到,因此必须选择充分有力的材料,让人信服地说明问题。

2. 适用范围

综译凭借其高度浓缩的信息,是介绍国外最新科研信息的重要手段。综译可以帮助各个学科引进国外最新研究发展成果,为国内学者提供某一学科发展的过去、现在、未来的发展动向。书写综译的时候要注意避免加入过多的主观评论,并且注意列出综译的参考文献以及注明引文的具体出处,以方便读者查找原文阅读。

3. 对译者要求

综译者必须熟练掌握运用外语,具备基本的翻译能力,并且对某一领域发展的过去、现状、未来趋势非常了解。可以说,综译者一般是某行业资深的信息专家,懂外文的专家,或在某一方面有深入研究的专家。

第三节　科技文体与翻译

一、翻译与文体

文体研究的一个主要目的是语言使用的得体性,而翻译的目的为译文的得体性。语言代码呈现出一定的文体特征,在翻译(即两种语言代码的转换)过程中,必然要重视相应的文体特征。实际上,得体与否正是衡量译品高下的尺度之一。

而翻译界的前辈已经对此有所论述。王佐良曾说:“似乎可以按照不同文体,定不同译法。例如信息类译意,文艺类译文,通知、广告类译体,等等。所谓意,是指内容、事实、数据等,须力求准确,表达法要符合当代国际习惯。所谓文,是指作家个人的感情色彩、文学手法、结构形式等,须力求保持原貌,因此常须直译。所谓体,是指格式、方式、措辞等,须力求符合该体在该语中的惯例……”

传统的文体学(stylistics)主要分析文学家的文学风格或作家作品的语言特点。而近半个世纪以来,应用性文体大大地发展并丰富了文体学的研究。

20 世纪 80 年代,随着系统功能语言学的发展,人们才从语言功能的角度把各种传递信息的语篇划归为实用文体,实用文体包含的语篇类型十分广泛,涉及社会生活、经济活动、科学技术、工农业生产、新闻传媒等方方面面,例如书信、函电、告示、契约、规章、报告、法律文件、旅游指南、广告、新闻报道、产品说明书、技术规范,等等。

文体学的研究在于:

(1) 分析各种语言习惯,以便确定哪些特征经常或仅仅应用于某些场合;

(2) 尽可能说明为什么某种文体具有这些特征,而不具备另一些特征;

(3) 以语言功能为依据,对这些特征进行分类。

二、科技语域的类别与层次

科技语域泛指一切论及科学和技术的书面语和口语,它的特点是层次多、范围广。

语境(context)的重要性早已为伦敦学派的马利诺夫斯基和弗斯所确认。韩礼德进一步明确了语言存在于语境(如会场、教室、语篇)之中。韩礼德把语境分解为语场(field)、语旨(tenor)和语式(mode)。语场指语言使用时所要表达的话题内容和活动,具体来说,即话语参与者正在从事的活动。语旨是语言使用者的社会角色和相互关系,以及交际意图。语式指进行交际所采用的信道、语篇的符号构成和修辞方式。

语境的这三大因素中任何一项的变化,都会引起交流意义的变化,从而引起语言的变异或文体的变化,产生不同类型的语域。据此,可将科技语域分为六个层次、两个类别。每个类别和层次的语言变体都有其语域特征:

文体	层次	正式程度	语场(题材或适用范围)	语旨(参与交际者)	语式(语言形式)
专用	A	最高	数学、力学等基础理论 科学论著、报告	科学家之间	语言成分主要为人工符号,用自然语言表示句法关系或略作说明
科技	B	很高	科学论著、法律文本(包括专利文件、技术标准、技术合同等)	高级管理人员之间、律师之间或专家之间	以自然语言为主,辅以人工符号
文体	C	较高	应用科学技术论文、报告、著作	同一领域的专家之间	
普通	D	中等	物质生产领域的操作规程、维修手册、安全条例等	生产部门的技术人员、职员、工人之间	自然语言,含部分专业术语,句法刻板
科技	E	较低	消费领域的产品说明书、使用手册、促销材料等	生产部门与消费者之间	自然语言,有少量术语,句法灵活
文体	F	低	科普读物、中小学教材	专家与外行之间	自然语言,避免术语,多用修辞格

下面,先初步比较同一内容的两类文体。

普通科技英语:

Food is digested as it passes along the long tube which begins at the mouth and ends at the anus. The process of digestion begins in the mouth where the food is crushed by the teeth and made wet by the juices in the mouth. After it is swallowed, it passes down the gullet, a tube having muscles and going to the stomach.

专用科技英语:

Food is digested as it passes along the alimentary canal. The process of digestion begins in the mouth where the food is chewed and moistened by the saliva. After being swallowed, it passes down the gullet, a muscular tube leading to the stomach.

普通科技英语	专用科技英语
1. the long tube which begins at the mouth and ends at the anus	the alimentary canal
2. is crushed by the teeth	is chewed
3. (is) made wet	(is) moistened
4. the juices in the mouth	the saliva
5. a tube having muscles	a muscular tube
6. going to the stomach	leading to the stomach

在专用科技英语中,常用拉丁语派生词(Latinate words)来替代同义的盎格鲁—撒克逊语单词(Anglo-Saxon words),以示高雅和正式,由于前者较后者词形长,常称为"big words"(大词),相反普通科技英语则以使用小词、短语为多。

在专用科技英语中常用规范的书面语动词来代替普通科技英语中的短语动词或动词词组。前者语义明确,后者有时多义,甚至易生歧义。

三、专用科技文体

首先,从语境角度来看专用科技文体。

语场:基础科学理论、技术性法律条文,涉及科学试验、科学技术研究、工程项目、生产制造等领域。

语旨:专家写给专家看的(expert-to-expert writing)。

语式:抽象程度很高的基础科学理论的论著,以人工语言为主,辅以自然语言,不是行家的翻译工作者难以理解。

专用科技文体具有以下特征:

1. 语义特征

语义特征主要为客观性,体现在:

(1) 语义结构显性化

专用科技文体由于正式程度高,逻辑严密、层次分明、条理清晰,在语义结构上排除歧义,语义关系表现在字面上。

在词汇衔接(lexical cohesion)方面,专用科技文体以同词重述为多,也用上义词替代。

The specimen can be inserted between pieces of similar hardness; the sample can be plated; when using casting resins, a slurry of resin and alumina made for just this purpose can be poured around the specimen; the specimen can be surrounded by shot, small revets, rings, etc., of about the same hardness.

译文:试样可夹在具有同样硬度的工件之间,可镀层。当用注塑树脂时,可将树脂与矾土的浆料注入试样周围,使试样被短小的护壁、环状填料(硬度差不多)等所包围。

在照应(reference),即语义所指方面,人称照应和指示照应均少使用。用得较多的是比较照应,比较照应是两个事物在比较中互相照应,使语句上下衔接。

(2) 非人称化

专用科技文体陈述客观规律和原理、反映客观现象和变化、描述客观过程和事实,突出"客观"两字,力戒主观意念和个人好恶。代词少用,即使使用,跟其他文体也有所区别。

但是近几十年来,也有人主张在学术论文或报告中适当使用人称代词,以加强学术交流中的亲和力,特别是阐发作者对科研工作的真实感情。

It has not escaped our notice that the specific pairing we have postulated immediately suggests a possible copying mechanism for the genetic material.

译文:引起我们注意的是:我们假定存在的特殊配对,直接使人联想到遗传物质可能的复制机理。

一般来说,为了表述客观,专用科技文体多用被动态。但是近二三十年来有多用主动态的倾向，认为主动态表达清晰、明确、亲切。

(3) 名词化结构的广泛使用

科技英语的文体特点是：行文简洁、表达客观、内容确切、信息量大、强调存在的事实，这些特点必然要求名词化现象的大量存在，而名词化的使用能很好地适合科技英语的一些特点。

The concept of internal resistance frequently provides an adequate description of batteries, generators, and other energy converters.

英文原句中使用了 V + N 结构，汉译中对句中的动词 provide 采用了省译的方法，并且将英文句中的名词 description 翻译为汉语中的动词。

译文：内阻的概念经常适当地描述了电池、发生器和其他能量转换器。

2. 用词正式

专用科技文体多用专业术语、大词、专用缩略语，辅以数学语言和工程图学语言，多为专家对专家的语言，外行往往不得要领。例如：

It is a further object of this invention to provide an apparatus which will have a proper weight to volume displacement to allow the thermocouple to sink beneath the surface of the molten ferrous bath in a basic oxygen furnace to thereby assure temperature measurements thereof.

译文：本发明的另一目的是提供一套装置，该装置以适当的重物作体积置换，而使热电偶沉入氧气顶吹转炉的熔池液面之下，从而保证其温度的测量。

3. 结构严密

在专用科技文体中，一方面大量采用缩略语、数学符号和工程图学语言使之简化，另一方面句型的扩展、连接手段的多样又使句法结构多层化。

词汇方面：

(1) 谓语动词常用一般现在时态，表示真理的普遍陈述，在时态上属于"零时态"(timeless)；

(2) 常用被动语态；

(3) 普遍使用名词词组及名词化结构；

(4) 宁用单个动词而不用短语动词。

句法方面：

(1)"无生命主语＋及物动词＋宾语(＋宾语补足语)"的句型比较常见；

(2) 常用 it 作形式主语或宾语；

(3) 宁用紧缩性状语从句而不用完整句；

(4) 割裂修饰比较普遍(包括短语或从句被割裂)；

(5) 句中并列成分(各种并列短语、单词或从句)较多;

(6) 句子长而复杂,句套句的情况较多。

Protection against complete structural failure is achieved in three different ways: first, by proper selection of material, especially in high load areas, to provide a consistent slow rate of crack propagation, and high residual strength; second, by providing such multipath structure on the airplane that the loss of any one segment would not endanger the airplane; and third, by providing readily accessible structure which can be inspected and maintained properly.

译文:用三种方法来防止整个结构的损坏:第一,适当选用材料,特别是在高载荷区的材料,要具有一致的、迟缓的裂痕扩散速率特性以及高剩余强度;第二,采用多路传力结构,使某一局部损坏不危及整个飞机;第三,使结构具有易卸性,便于检查与维修。

程式化:

程式化是指同类语篇大致相同的体例和表达方式。例如,期刊论文的体例按次为:① Title(标题);② Abstract(摘要);③ Introduction(引言);④ Materials and Methods(材料与方法)或 Equipment and Test/Experiment Procedure(设备与试验/实验过程);⑤ Result(结果);⑥ Discussion(讨论);⑦ Summary 结论(概要);⑧ Acknowledgments(致谢);⑨ References(参考文献)。

四、普通科技文体

普通科技文体一般涵盖科普文章(popular science/science article)和技术文本(technical prose/document)。

首先,从语境角度来看普通科技文体。

语场:传播科技知识、描写生产过程、说明产品的使用方法等。

语旨:内行对外行(scientist/journalist-to-layperson writing)。

语式:采用自然语言,偶用人工符号,用词生动,句法简易,文风活泼,多用修辞格。

技术文本主要有:

(1) 通信;

(2) 营销;

(3) 产品操作指南和使用说明书;

(4) 建议书;

(5) 报告书。

普通科技文体常以简明的语言、生动的方式把信息传递给受众。与专用科技文体不同,普通科技文体的主要特点是:用词平易、句式简单、多用修辞格。

普通科技文体具有以下特征:

1. 用词平易

用常用词代替专业术语：

Scientists have made maps showing the "earthquake belts".

A scientist dived below the surface of the sea in a hollow steel ball.

用动词代替抽象名词：

在普通科技文体中，少用名词化结构。

We ask that all employees cooperate. (We ask for the cooperation of all employees.)

The property was acquired through long and hard negotiations. (The acquisition of the property was accomplished through long and hard negotiations.)

多用代词：

使用人称代词可拉近作者与读者的距离，有亲切感。

When you choose the Shut Down command from the Start Menu, you see a dialog box that asks, "Hey, what do you mean, 'shutdown' "?

2. 句法简单

使用短句：

与专用科技文体不同，普通科技文体多用短句，以使文字简洁有力，生动活泼。一般而言，短句语义明晰，通俗易懂。

避免使用形式主语 it 和引导词 there：

光线从太阳到地球要 8 分钟。

译文一：It takes about 8 minutes for light to reach the earth from the sun.

译文二：Light takes about 8 minutes to reach the earth from the sun.

一号矿井有爆炸的危险。

译文一：There is a danger of explosion in Number One mine.

译文二：Number One mine is in danger of explosion.

多用主动态：

在普通科技文体中，"主谓宾"的句型是最基本的句型，它有利于作者与读者的交流，创造一种非正式的语气。例如：

In order to complete this chapter and our overview of the Internet, we need to spend a few moments talking about TCP/IP. As you know, the Internet is built on a collection of network that cover the world. These networks contain many different types of computers, and somehow, something must hold the whole thing together. That something is TCP/IP.

3. 多用修辞格

现代功能修辞学从传统的文学修辞研究发展到对不同文体的修辞研究。

对于翻译而言,处理好科技文体中的修辞格与处理文学中的修辞格同样具有重要的意义。

Like a solidly hit fast ball, the Big Bang is going-going-going.

译文:像一个被猛击一下飞速远去的棒球,大爆炸正渐行渐远,一去不复返。

The program ready for computer to "read" is prepared in a specially designed language.

译文:准备给计算机"读"的程序,是用专门设计的语言编制的。

参考答案

Unit 1　Mathematics

Text A　Mathematics

Ⅰ. 1. F　2. T　3. T　4. F　5. F　6. F

Ⅱ. 1. Practical mathematics has been a human activity for as far back as written records exist.

2. Experimental mathematics continues to grow in importance within mathematics because computation and simulation are playing an increasing role in both the sciences and mathematics.

3. In the 18th and 19th centuries, the industrial revolution led to an enormous increase in urban populations. Basic numeracy skills, such as the ability to tell the time, count money and carry out simple arithmetic, became essential in this new urban lifestyle.

4. Fields Medal.

5. At least nine of the problems have now been solved.

Ⅲ. 1. 数学证明　2. 应用数学　3. 计算模拟;运算模拟　4. 英国国家课程标准　5. 标准化数学考试　6. game theory　7. theoretical physics　8. elementary mathematics; fundamental mathematics　9. multiplication and division　10. pure mathematics

Ⅳ. 对于许多学科领域数学都是基础,这些学科领域包括自然科学、工程、医学、金融和社会科学。应用数学促使全新数学相关专业领域产生,比如统计学和博弈论。数学家当然也从事理论数学的研究,也就是研究数学自身,不考虑数学的应用。理论数学和应用数学之间没有明显的界线,经常发现一些开始目的是理论数学的研究会带来实际的应用。

Text B　Applied Mathematics

1. Applied mathematics is a branch of mathematics that deals with mathematical methods that find use in science, engineering, business, computer science, and industry.

2. Applied analysis, applied analysis, differential equations and approximation theory of mathematics related directly to the development of Newtonian physics.

3. It includes the classical areas noted above as well as other areas that have become increasingly important in applications. Even fields such as number theory that are part of pure mathematics are now important in applications.

4. The advent of the computer has enabled new applications: studying and using the new computer technology itself (computer science) to study problems arising in other areas of science (computational science) as well as the mathematics of computation.

5. Mathematical economics is the application mathematical methods to represent theories and

analyze problems in economics. The applied methods usually refer to nontrivial mathematical techniques or approaches. Mathematical economics is based on statistics, probability, mathematical programming (as well as other computational methods), operations research, game theory, and some methods from mathematical analysis.

Unit 2 Physics

Text A The Science of Matter, Space and Time

Ⅰ. 1. F 2. F 3. F 4. T 5. F 6. F

Ⅱ. 1. The six quarks and six leptons.

2. Theorists speculate that there may be other types of building blocks, which may partly account for the dark matter.

3. Four elementary types of forces act among particles: strong, weak, electromagnetic and gravitational force.

4. Particles transmit forces each other by exchanging force-carrying particles called bosons.

5. Because, at the subatomic level, the gravitational force is many orders of magnitude weaker than the other three elementary forces.

Ⅲ. 1. 粒子物理学 2. 亚原子类 3. 粒子与反粒子 4. 量子效应 5. 质子/中子/分子 6. matter and antimatter 7. astrophysics 8. gravitational force 9. nucleus 10. graviton

Ⅳ. 通过交换称为玻色子的携带玻色子的粒子,粒子相互传递作用力,这些作用力的介质从一个粒子到另一个粒子携带离散量的能量,称为量子。您可以将玻色子交换引起的能量转移想象成两个篮球球员之间在传球。物理学家估计,引力也可能与名为引力子的玻色子有关。这个假想的玻色子极难观察,因为在亚原子水平上,引力比其他三个基本作用力弱了许多个数量级。

Text B End of the Smashed Phone Screen? Self-healing Glass Discovered by Accident

1. Japanese researchers say they have developed a new type of polymer glass that can mend itself.

2. The new material was the first hard substance of its kind that can be healed at room temperature, while self-healing rubber, self-healing plastics and healable glass need high heat to melt the material.

3. The research was led by Professor Takuzo Aida from the University of Tokyo.

4. No. The research promises healable glass that could potentially be used in phone screens and other fragile devices.

5. According to the over 21% of UK smartphone users were living with a broken screen, with smashed displays being one of the biggest issues alongside poor battery life.

Unit 3 Chemistry

Text A What Is Chemistry?

Ⅰ. 1. F 2. F 3. F 4. T 5. F 6. T

Ⅱ. 1. Chemistry is involved in everything we do. Chemistry is one of the physical sciences that help us to describe and explain our world.

2. They are analytical chemistry, physical chemistry, organic chemistry, inorganic chemistry and biochemistry.

3. Carbohydrates are sugars and starches, the chemical fuels needed for our cells to function.

4. Lipids are fats and oils. Because fats have 2.25 times the energy per gram than either carbohydrates or proteins, many people try to limit their intake to avoid becoming overweight. However fats and oils are essential parts of cell membranes and to lubricate and cushion organs within the body.

5. These characteristics are dissolved oxygen, salinity, turbidity, suspended sediments, and pH.

Ⅲ. 1. 自然科学 2. 定性和定量观测 3. 悬浮沉积物 4. 营养补充;营养补充剂 5. 脂类;脂质 6. carbohydrate 7. water pollution; water contamination 8. chemical reaction 9. chemical bond 10. organism

Ⅳ. 地球化学家结合化学和地质学,研究在地球上发现的物质之间的组成和相互作用。与其他类型的化学家相比,地球化学家在实地研究上可能花费更多的时间。美国地质调查局或环境保护局做了许多工作,以确定采矿作业和废物如何影响水质和环境。他们可能前往偏远的废弃矿场采集样品并进行粗略现场评估,然后沿着一条小溪穿过其流域,评估污染物是如何通过系统的。

Text B New Clues Could Help Scientists Harness the Power of Photosynthesis

1. The origin of life, the invention of DNA and photosynthesis.

2. It absorbs light in the far-red range of the light spectrum.

3. Until now, there were few known intermediate stages in its evolution.

4. The enzyme requires light to catalyze its reaction and may not require oxygen.

5. Because the ability confers a powerful advantage to those organisms that produce chlorophyll *f*—they can survive and grow when the visible light they normally use is blocked.

Unit 4 Mechanical Engineering

Text A Mechanical Engineering Overview

Ⅰ. 1. T 2. T 3. F 4. F 5. F 6. F

Ⅱ. 1. The career paths of mechanical engineers are largely determined by individual choices, a decided advantage in a changing world.

2. Mechanics, energy and heat, mathematics, engineering sciences, design and manufacturing.

3. Mechanical engineers research, design, develop, manufacture, and test tools, engines, machines, and other mechanical devices.

4. The first step is to visualize what is happening and clearly state the problem.

5. Design is one of the most satisfying jobs for a mechanical engineer. Synthesis is when you pull all the factors together in a design that can be successfully manufactured. Design problems are challenging because most are open-ended, without a single or best answer. There is no best mousetrap — just better ones.

Ⅲ. 1. 工程学科 2. 机床 3. 可视化技术 4. 化石燃料 5. 替代来源 6. generator 7. enterprise management; business administration 8. internal combustion engine 9. gas turbine 10. ventilation system

Ⅳ. 从根本上说，机械工程师从事研究运动力学以及能量从一种形式到另一种形式或从一个地方到另一个地方的转移。机械工程师设计和制造用于工业和消费类的机器——实际上，几乎任何机器，都有机械工程师参与其开发和生产。他们设计供暖、通风和空调系统，以控制家庭、办公室和工厂的温度，并开发食品工业的制冷系统。机械工程师还设计热交换器——高科技机械和电子计算机设备中的关键组件。

Text B Mechanical Engineering

1. Computer-aided design, and product lifecycle management.

2. Mechanical engineering emerged as a field during the Industrial Revolution in Europe in the 18th century.

3. Al-Jazari, who was one of them, wrote his famous *Book of Knowledge of Ingenious Mechanical Devices*④ in 1206, and presented many mechanical designs. He is also considered to be the inventor of such mechanical devices which now form the very basic of mechanisms, such as the crankshaft and camshaft.

4. Sir Isaac Newton both formulated the three Newton's Laws of Motion and developed Calculus, the mathematical basis of physics. Gottfried Wilhelm Leibniz⑥ is also credited with creating Calculus during the same time frame.

5. Early 19th century.

Unit 5 Electrical and Electronic Engineering

Text A Electrical Engineering

Ⅰ. 1. F 2. T 3. F 4. T 5. F 6. F

Ⅱ. **1**. The invention of the transistor, and later the integrated circuit.

2. Electrical engineering includes a wide range of subfields such as electronics, digital computers, power engineering, telecommunications, control systems, radio-frequency engineering, signal processing, instrumentation, and microelectronics. The subject of electronic engineering is often treated as its own subfield but it intersects with all the other subfields, including the power electronics of power engineering.

3. The invention of the transistor in late 1947 by William B. Shockley, John Bardeen, and Walter Brattain of the Bell Telephone Laboratories opened the door for more compact devices and led to the development of the integrated circuit in 1958 by Jack Kilby and independently in 1959 by Robert Noyce. Starting in 1968, Ted Hoff and a team at the Intel Corporation invented the first commercial microprocessor, which foreshadowed the personal computer.

4. Control engineering has a wide range of applications from the flight and propulsion systems of commercial airliners to the cruise control present in many modern automobiles. Control systems play a critical role in space flight. It also plays an important role in industrial automation.

5. Microelectronics engineering deals with the design and microfabrication of very small electronic circuit components for use in an integrated circuit or sometimes for use on their own as a general electronic component.

Ⅲ. **1**. 电气工程 **2**. 电子工程 **3**. 控制系统 **4**. 国际工程技术学会 **5**. 电位差 **6**. telegram **7**. integrated circuit **8**. direct current **9**. electromagnetic radiation **10**. microprocessor

Ⅳ. 微电子工程学涉及非常小的电子电路组件的设计和微制造,这些组件用于集成电路,有时也可以单独用作一般的电子组件。尽管可以在微观层级上创建所有主要的电子组件(电阻、电容器等),但最常见的微电子组件是半导体晶体管。纳米电子学将器件进一步缩小到纳米级。现代设备已经处于纳米状态,自2002年以来,低于100纳米的处理已成为标准。

Text B Basic Electronics Concepts and Theory

1. No. The flow of electrons is the reverse.

2. Any current flowing in a wire will create a magnetic field.

3. Resistors are one of the most common passive electrical components and are found in almost every electronic circuit. Heating effect is a common example of resistor application.

4. Electrical resistance is measured in ohms and can be determined through the use of Ohms Law.

5. The amount of steady current through a large number of materials is directly proportional to the potential difference, or voltage, across the materials.

$I = V/R$

I = current (amps)

V = voltage (volts)

R = resistance (ohms)

Unit 6 Agriculture

Text A Labor Mobility and the Rural Exodus

Ⅰ. 1. T 2. F 3. F 4. T 5. T 6. F

Ⅱ. 1. The earlier, which is over a hundred years old, is from the savannah interior to export-cropping zones nearer the coast, notably the forest belt. This rural-rural migration pattern is still important, but extremely difficult to quantify, because rates of natural population increase also vary widely within the region. The second movement, which is substantially a feature of post-World War II history, is the rapid growth of West African cities, especially capital cities.

2. There are two main ways in which it could be said that commercial agriculture has contributed to rural emigration.

3. Because of the mobility in the period of extraordinary growth.

4. Work conditions are determined by what keeps the workers happy.

5. When people are concentrated in central places, markets are bigger and services can be delivered more cheaply on a per capita basis.

Ⅲ. 1. 受约束的网络 2. 商业性农业 3. 人口结构变化 4. 农村社会结构 5. 农业盈余 6. rural population 7. natural population growth 8. material amenities 9. rules of exogamy 10. export-crop agriculture

Ⅳ. 事实上,城市的物质设施的质量和数量都比大多数乡村的更好更多。此外,当人们集中在中心地区时,市场就会更大,按人均计算,提供服务的成本也会更低。乡村生活有其坚实的优点,但那里的经济视野必然是有限的,而这种有限关系到很多人,尤其是年轻人。国家统治者的经济行为将成千上万的村民吸引到他们的首都城市,他们有时可能会后悔自己如此靠近权力走廊,如果他们感到不安,就会在那里造成更大的破坏。

Text B The Traditional Organization of Farming

1. The effects of these developments on the rural division of labor as a whole.

2. Because many people lived in or near towns and cities where they earned their livelihoods principally from manufacturing and services.

3. The basic tool remained the hoe: There was no use made of the plow or irrigation. Animal traction was more or less absent, except for the use of donkeys for transport in some area.

4. In the more remote parts of the savannah and forest, sex, age, and kinship were the sole determinants of the division of labor.

5. Men rule, farm, dye, build, work metals, skin, tan and work leather, slay and handle cattle and small livestock, sew all sewn clothes, make musical instruments and music, trade, keep bees, weave mats, may be Mallams (teacher-priests), wash clothes, weave narrow cloth on the men's loom, go on long-distance trading expeditions, make pots, do carpentry—native and

European—are the doctors and magicians, the barbers, employed farm laborers, brokers and taxpayers. They also fish, hunt and do all the family marketing, keep goats, sheep, chickens, ducks, turkeys and pigeons, and take part in war.

Unit 7 Plants and Animals

Text A Early Human Societies and Their Plants

Ⅰ. **1.** F **2.** T **3.** T **4.** F **5.** T **6.** T

Ⅱ. **1.** This notion of a sudden agricultural revolution originated because of what appeared to be the almost overnight appearance and cultivation of new forms of several key plants, especially cereals and pulses, that had supposedly been deliberately "domesticated" by people.

2. No, it is not right to say so.

3. Over the past decade or so, detailed archaeological and genetic evidence has emerged supporting the view that widespread cultivation of crops evolved separately in various parts of Asia, Africa, Mesoamerica, and South America.

4. In these latter three regions, crops and agronomic techniques were only secondarily acquired from the primary agricultural societies. These crops were then grown in places that were far from their initial centres of origin.

5. The second most popular class of staple domesticants were the starchy tubers such as yams and potatoes.

Ⅲ. **1.** 农业革命 **2.** 开发的种植物 **3.** 考古学和遗传学证据 **4.** 特色农业体系 **5.** 耕作前的阶段 **6.** human culture **7.** urbanization **8.** edible plant **9.** staple crops **10.** defining moment

Ⅳ. 在过去的十多年里,详细的考古学和遗传学证据已经出现,支持了这样一种观点,即广泛种植的农作物在亚洲、非洲、中美洲和南美洲的不同地区是分别进化而来。相比之下,在欧洲、北美和澳大拉西亚,作物种植发生得要晚得多。在后三个地区,作物和农艺技术只能从初级农业社会获得。然后,这些作物被种植在远离最初原产地的地方。

Text B The Traveling Exotic Animal Protection Act

1. The Traveling Exotic Animal Protection Act, a. k. a. TEAPA, would restrict the use of exotic animals in traveling circuses in the United States.

2. TEAPA Amends S. 13 of the Animal Welfare Act to the effect that no exhibitor may allow the (use) of an exotic or wild animal (including a non-human primate) in an animal act if it is living in mobile accommodation and is constantly traveling (the United States).

3. The bill does not apply to zoos, aquariums, universities, laboratories, rodeos, horse racetracks or facilities where the animals reside permanently and are not constantly traveling. The bill also does not apply to domestic animals.

4. From cruel training techniques to the misery of confinement, there are many abuses that animals in circuses suffer.

5. Urge your federal legislators to support the Traveling Exotic Animal Protection Act.

Unit 8 Computer Science

Text A Supplemental Skills for Success in 3D

Ⅰ. 1. T 2. F 3. T 4. F 5. T 6. T

Ⅱ. 1. If you aspire to success as a 3D artist, the best way to improve is to practice as much as possible.

2. Life drawing is often seen as the "granddaddy" of all skills in the artistic world, because if you're truly great at it you're pretty much worth your weight in gold.

3. James Gurney, who wrote and illustrated the *Dinotopia* series.

4. After some time behind a lens you'll start to learn what works and what doesn't, and what sort of lighting produces compelling imagery. Another advantage of picking up photography (or studying good photographs)—it's probably the fastest and easiest way to learn composition.

5. If you really want to become a great animator, do yourself a favor and act!

Ⅲ. 1. 人体素描 2. 竞争激烈的行业 3. 无可争议的大师 4. 色盘 5. 内在把握 6. figurative sculpture 7. character artist 8. sculpting workflow 9. modeling 10. environment artist

Ⅳ. 随便问任何一个专业的摄影师——都是光线问题。正如它所发生的那样，光线在计算机图形界也是非常重要的——如果你没有一个好的光线解决方案，世界上最好的模型也不会产生一个好的渲染效果。如果你是一个有抱负的3D艺术家，或者你认为灯光是你工作流程中的弱项，试着拿起相机做一些摄影。在镜头后面一段时间后，你将开始了解什么可行，什么不可行，以及什么样的光线可以产生引人注目的图像。

Text B Cyberterrorism: Latest Threat to National Computer Security?

1. Invading cyberspace and networks as a form of terrorist attack.

2. A number of experts conclude that while the threat of cyberterrorism does exist, it is routinely exaggerated by various actors for political or financial gain.

3. The Patriot Act and new offices address cyberterrorism.

4. To counter the threat of cyberterrorism by expanding the punishments for those committing cybercrimes, and pushing for Internet Service Providers to disclose information to the government for investigations.

5. The aim of the National Cyber Security Division is to build a national cyberspace response system, and to put into use a cyber-risk management program to protect critical infrastructure.

Unit 9 Information and Communication Technology

Text A An Introduction to Information Technology (IT)

Ⅰ. **1.** T **2.** T **3.** F **4.** F **5.** T **6.** F

Ⅱ. **1.** A 1958 article in *Harvard Business Review*① referred to information technology as consisting of three basic parts: computational data processing, decision support, and business software.

2. The challenges are data overload, teamwork and communication, system and network security.

3. To manage the computer technologies related to their business.

4. Because networks play a central role in the operation of many companies, business computer networking topics tend to be closely associated with Information Technology.

5. Having success in this job field requires a combination of both technical and business skills.

Ⅲ. **1.** 信息技术 **2.** 计算数据处理 **3.** 数据库管理 **4.** 商业软件部署 **5.** 信息安全 **6.** software development **7.** computer system architecture **8.** project management **9.** cloud service **10.** computation model

Ⅳ. 随着各种计算系统和计算能力在全球范围内的不断扩展,数据过载已成为许多专业 IT 人员日益关注的问题。高效地处理大量数据以产生有用的商业智能需要大量的处理能力、复杂的软件和人工分析技能。

Text B Parent Connection Goals Are Best Served by Technology

1. The National Parent Teachers Association (PTA) ① Standards for Family-School Partnerships suggests that "families, the community, and school staff communicate in numerous interactive ways, both formally and informally."

2. When is a technology best suited for general communication between parents and teachers?

What is best platform for widespread news/message dissemination to all stakeholders?

Which software/apps are best suited for private communication about a specific student?

What technologies are already being used effectively by both parents and teachers?

What types of hardware do parents and teachers have access to on a regular basis? (computers, smartphones, land-lines, etc.)

What kinds of skills do parents have with technology and to what extent are they using it?

What kinds of skills do educators have with technology and to what extent are they currently using those skills?

How can face-to-face communication between parents and educators be supported (not replaced) with technology?

How can teachers in different disciplines (Gr 7－12) collaborate and coordinate separate communication using technology?

Who is the "gatekeeper" for outgoing communication in a 7－12 school?

3. Time is the best reason for educators to design communication using technology.

4. For educators, schools now have databases for e-mail addresses, for robo-calls and for text messages in order to make parent contact.

5. Language barriers, equal access to information, etc.

Unit 10 Energy Science

Text A All Types of Coal Are Not Created Equal

Ⅰ. 1. T **2.** T **3.** F **4.** F **5.** T **6.** T

Ⅱ. 1. People don't "produce" coal. Geological processes and decaying organic matter create it over thousands of years.

2. Australia tops the worldwide list of exporters, having sent 298 million metric tons of coal overseas in 2010.

3. Coal falls into two main categories: hard and soft.

4. The stored energy potential within coal is described as the "calorific value", "heating value" or "heat content."

5. As a general rule, the harder the coal, the higher its energy value and rank.

Ⅲ. 1. 燃烧化石燃料 **2.** 地质作用 **3.** 炼焦煤 **4.** 动力煤 **5.** 褐煤 **6.** volatile compound **7.** impurities **8.** iron ore **9.** physical properties **10.** unit of heat

Ⅳ. 拥有不同组成成分的煤在世界各地被用作发电和炼钢的可燃矿物燃料。据国际能源机构称,它是21世纪全球增长最快的能源。人们不"生产"煤。地质作用和腐朽的有机物在数千年中创造了它。它是从地下构造或"煤层"中开采出来的,通过地下隧道,或者通过移除地表的大面积区域。挖出的煤必须经过清理、洗涤和加工,以备商业使用。

Text B Energy for Future Presidents

1. This section has 4 chapters focusing on the important sources of power production with current. energy policy and technologies.

2. *Physics for Future Presidents*.

3. Muller divides his book into 5 distinct sections, which each focuses on different elements of the current energy debate in our country.

4. Chapter 16: Electric Automobiles

5. Yes, he is a scientist.

Unit 11 Petrochemistry

Text A Crude Oil

Ⅰ. 1. F 2. T 3. T 4. F 5. T

Ⅱ. 1. Crude oils are compounds that mainly consist of many different hydrocarbon compounds that vary in appearance and composition.

2. Average crude oil composition is 84% carbon, 14% hydrogen, 1%–3% sulphur, and less than 1% each of nitrogen, oxygen, metals and salts.

3. Crude oils are distinguished as sweet or sour, depending upon the sulphur content present. Crude oils with a high sulphur content, which may be in the form hydrogen sulphides, are called sour, and those with less sulphur are called sweet.

4. Refining is the processing of one complex mixture of hydrocarbons into a number of other complex mixtures of hydrocarbons.

5. One of the major concerning issues in today's world is the dependence of the modern society on oil and gas and various other petroleum products.

Ⅲ. 1. 原油 2. 碳氢化合物 3. 硫黄含量 4. 氢硫化物 5. 分馏 6. explosive vapors 7. by-product 8. manufacturing process 9. crude oil raw materials 10. complex distillation

Ⅳ. 直接从油井中提取的未加工或未加工处理的(“原油”)是无用的。虽然轻质低硫原油被直接用作燃烧燃料,但这些较轻的碎片在燃料箱中形成爆炸性蒸气,因此是危险的。在用作燃料和润滑剂之前,以及在一些副产品形成材料[如塑料、洗涤剂、溶剂、弹性体和纤维(如尼龙和聚酯)]之前,必须将油分离成不同的部分并进行精制。

Text B Petrochemistry

1. Yes. Petroleum, produced over millions of years by natural changes in organic materials, accumulates beneath the earth's surface in extremely large quantities.

2. The demand for synthetic materials increased, and this rising demand was met by replacing costly and sometimes less efficient products with these synthetic materials. This caused petrochemical processing to develop into a major industry.

3. The first oil commercial was set up in 1859, two years after which the first oil refinery was set up.

4. The industry used basic materials: synthetic rubbers in the 1900s, Bakelite, the first petrochemical-derived plastic in 1907, the first petrochemical solvents in the 1920s, polystyrene in the 1930s.

5. Yes. Petrochemistry is a science that can readily be applied to fundamental human needs, such as health, hygiene, housing and food.

Unit 12 Aeronautics

Text A NASA's Chief Scientist: The Future of Space Exploration Is International Partnerships

Ⅰ. 1. F 2. F 3. F 4. T 5. T 6. T

Ⅱ. 1. Because some believe Mars is the planet that most resembles the Earth, some think Mars has loomed large in the public's imagination, and the others see it as inspirational and good for their countries' economy.

2. Because NASA's view is to turn over to the private sector those projects that in a sense have become routine so it can focus its resources on getting to Mars.

3. How to get to Mars and back again safely.

4. The latter is more global and involve greater private and public partnerships.

5. The US and Soviet Union.

Ⅲ. 1. 太空探索 2. 阿波罗计划 3. 太空争霸 4. 联邦预算 5. 火星人 6. The European Space Agency 7. asteroids 8. space flight 9. Mars 10. science fiction

Ⅳ. 同样明显的是,下一阶段的太空探索不仅将更加全球化,而且将涉及更多的私人和公共伙伴关系。像 Space X 和波音这样的公司正越来越多地参与到美国国家航空航天局的日常行动中,其中包括一个联合项目,该项目可能在 2017 年将宇航员送上太空。美国国家航空航天局的观点是,将那些在某种意义上已经成为常规的项目移交给私营部门,斯托芬建议,让美国国家航空航天局将其资源集中在到达火星上。

Text B A "Starshade" Could Help NASA Find Other Earths Decades Ahead of Schedule

1. They can use it to take snapshots of "other Earths" to study worlds beyond our solar system for signs of habitability and life.

2. First it's about light years distance, a habitable world would be a faint dot lost in the overpowering glare of its larger, 10-billion-times-brighter star. Second it's the Earth's turbulent, starlight-blurring atmosphere.

3. ① A starshade is a device of sunflower-shaped, paper-thin screen half as big as a football field, which can float tens of thousands of kilometers directly ahead of WFIRST, blocking out a target star's light with an extended thumb.

② Initially folded up for launch into space, the starshade would detach and unfurl and then fly away to its station tens of thousands of kilometers ahead of the telescope.

4. More than 3,000.

5. On a test bed, which is a meter-wide, 75-meter-long tube with a camera at one end, a laser at the other and a scaled down starshade in between.

Unit 13 Auto Industry

Text A The Major Problem with Cheap Electric Cars

Ⅰ. **1**. F **2**. T **3**. F **4**. T **5**. T **6**. F

Ⅱ. **1**. Because it lacks the sex appeal of Tesla, it's not nearly as practical as the plug-in Chevy Volt.

2. Because Federal tax credits and other incentives are used to bring down the price of electric vehicles to stimulate their sales.

3. 15%.

4. Nissan and Chevrolet.

5. Even at that low price, it's not clear if all that many drivers will buy them.

Ⅲ. **1**. 电动汽车 **2**. 标价 **3**. 销售记录 **4**. 封闭的高尔夫球车 **5**. 不常使用的汽车 **6**. light vehicles **7**. plug-in vehicles **8**. net takeaway price **9**. pure-electric cars **10**. Federal tax credits

Ⅳ. 希望降价能使三菱公司电动汽车的月销量超过12辆。但即使以这个价格,众多开车人是否会买账,仍不得而知。《消费者报告》杂志的评论说,i-MiEV电动汽车"不是一款能让人乐在其中的汽车",而且,该车的最新折扣价"对于一款与封闭的高尔夫球车没什么两样的车来说,还是太贵了。它的吸引力仅仅在于能使客户获得纯电动汽车的驾驶体验而已。就这个价位来讲,把它定位为不常使用的备用车似乎更可行"。

Text B The Dangers of an Exploding Car Battery

1. The basic idea behind safely connecting jumper cables is to connect the electrical system of a donor vehicle, which has a good battery, to the electrical system of a vehicle with a dead battery.

2. Because they utilize plates of lead submerged in sulfuric acid to store and release electrical energy.

3. They aren't a terribly efficient way to store energy and they're made up of fairly hazardous materials which can interact in dangerous ways.

4. One is a spark created when connecting or disconnecting a jumper or charging cable and another is the hydrogen gas present in the cell.

5. There are two main types of sealed car batteries: traditional lead acid batteries and VRLA (valve-regulated lead acid) batteries.

Unit 14 Communication and Transportation

Text A Apple Watch Retrospective: One Year Later

Ⅰ. **1**. T **2**. F **3**. T **4**. T **5**. F **6**. T

Ⅱ. 1. The Apple Watch, Apple Watch Sport, and Apple Watch Edition.

2. Allow the support of native apps on the Watch, new watch faces including a 24-hour timelapse options, a nightstand mode, the ability to draw in multiple colors, and improvements to the Maps and Apple Pay app.

3. A number of watch bands, seasonal lineups of its traditional bands with new color choices, and a charge cable, charging docks and cases.

4. The Hermès versions of the Apple Watch come sporting an exclusive Hermès watch face, and exclusive watch bands unique to the collection.

5. A software and hardware hack.

Ⅲ. 1. 价格标签;标价 2. 软件更新 3. 床头钟模式 4. 尼龙表带 5. 心率检测仪器 6. cellular data 7. accelerometer 8. charging dock 9. third-party manufacturers 10. boot up

Ⅳ. 上周的传闻还指出了支持蜂窝数据的下一个版本 Apple Watch,该功能让您在不携带 iPhone 的情况下也可使用 Apple Watch。目前分析师预计,苹果将在 5 月的 WWDC 上正式宣布该设备的下一个版本。最近调查表明,如果有升级的话,大多数 Apple Watch 持有者都会对升级感兴趣。

Text B High Speed Trains

1. At speeds of around 135 mph (217 km/h).

2. No. Because other technologies are easier to implement and they allow for more direct high speed connections to cities without the need for new tracks.

3. France.

4. They have the potential to relieve congestion on other transit systems.

5. First, high speed trains are considered more energy efficient or equivalent to other modes of transit per passenger mile. Second, they can reduce the amount of land used per passenger when compared to cars on roads.

Unit 15 Digital Technology

Text A Digital Spies

Ⅰ. 1. F 2. T 3. F 4. T 5. T 6. F

Ⅱ. 1. Future superpowers will be those nations with the greatest capability to harness the power of the electron for both economic and "digital" warfare.

2. The traditional roles of spies in gathering, communicating and analyzing information (secrets), as well as counterintelligence, have been greatly altered.

3. The transformation of the Internet into the information highway. By high-speed Internet access, advanced networking to share information quickly, and massive computer power to analyze

billions of bits of data to discover the secrets hidden inside.

4. ① Messages, information and signals are now transmitted in ways that appear innocuous but almost defy detection.

② Communication between a spy and his handler may occur in milliseconds.

③ Advanced encryption techniques may be utilized to additionally mask data.

5. Computer viruses can be developed and deployed that will be activated in a hostile opponent's computer to disable its computer infrastructure and cripple its economy, communications and defense.

Ⅲ. **1.** 情报局 **2.** 反情报机构 **3.** 先进感应镜片 **4.** 语音识别软件 **5.** 加密技术 **6.** digital scanner **7.** decipher **8.** meteorological satellite **9.** artificial intelligence **10.** information highway

Ⅳ. 互联网到信息高速公路的转变永远改变了信息采集的方式。间谍需要获取的信息中,大约90%是公开的。互联网作为世界知识的图书馆,已经成为世界超级大国所需的信息宝库。获得这些知识的关键是高速互联网访问,快速共享信息的高级网络以及强大的计算机功能,可以分析数十亿位的数据来发现隐藏在其中的秘密。在不久的将来,一个聪明的计算机程序员在一天内恢复的信息,要比一千个虚构的 James Bonds 在一生中能够恢复的重要信息多得多。

Text B　How to Take Advantage of Mobile Apps for Monitoring Your Health

1. Device interoperability.

2. To digitalize fitness data from disparate exercise equipment through pattern recognition and then coalesce this information so it becomes more useful for the user.

3. It is a mobile health technology that can transfer real-time data from Non-Connected Devices to Health IT Systems.

4. ① They have a lot of limitations in terms of storing and sharing information in a meaningful way.

② They are often not practical nor are they compatible with modern data collection practices—let alone the tempo of modern life.

5. First, they are not only more effective but also more enjoyable and compatible with other daily activities.

Second, this innovative digital technology is aiming to bridge the gap between patients who have difficulty traveling but still would like to share data with their health-care provider, creating a win-win situation.

Unit 16　Ecology

Text A　Ecology of Transgenic Crops

Ⅰ. **1.** T **2.** F **3.** T **4.** T **5.** T **6.** F

Ⅱ. 1. The advantages include increased yields, improved flavor or nutritional quality of foods and reduced pesticide use.

2. Harmful effects on human health and environment.

3. The investigator must summarize a large collection of studies across many species and situations, which can be done with a statistical approach called meta-analysis.

4. No. Because it would require many experiments: testing the transgenic plant in different environmental conditions, at different times of year, in combination with different farming practices, and examining the effects of the plants and plant by-products on an enormous number of species that could potentially be affected by the transgenic traits. However, attaining this level of certainty is neither reasonable nor possible.

5. Objective.

Ⅲ. 1. 商业种植 2. 基因工程 3. 杀虫剂 4. 食草动物抗性 5. 引进植物;引种植物 6. transgenic crops 7. gene 8. bacterium (pl. bacteria) 9. hybridization 10. breeding

Ⅳ. 同样,随着时间的推移,转基因作物可能导致重大问题。到目前为止,很少有实验研究转基因作物的安全性,尤其是这些改变对环境的影响。此外,生产和销售转基因作物的公司有责任评估转基因作物的安全性,这可能会导致潜在的利益冲突,影响安全性评估的严谨性。最后,由于某些风险来源于罕见的偶然事件,包括作物和杂草亲属之间的杂交,所以可能需要很长时间才能出现问题。同时,转基因作物面世以后,用于环境监测的资金和时间有限,因此,有可能在问题发生很久之后,都没有人会发现任何问题的迹象。

Text B Keystone Species and Their Role in Ecology

1. A keystone species is a plant or animal that plays a vital role in the health of the ecosystem in which it lives.

2. The term "keystone species", was first used in 1969 by University of Washington Zoology professor Robert T. Paine in his studies of the Makah Bay in Washington.

3. The key to understanding the role of predators is understanding the effect their disappearance would have on the health of the ecosystem.

4. The mutualists keystone species is the one whose actions benefit others in the ecosystem in such a way that without them, other species would not survive.

5. The loss of a keystone would make it impossible for other species in a habitat to survive.

Unit 17 Oceanography

Text A An Introduction to Oceanography

Ⅰ. 1. F 2. F 3. T 4. F 5. T 6. F

Ⅱ. 1. Oceanography is a discipline within the field of Earth sciences (like geography) that is focused entirely on the ocean.

2. Recent oceanographic studies have involved the use of modern technology to gain a more in depth understanding of the world's oceans.

3. Oceanography is multi-disciplinary and incorporates a number of different sub-categories or topics, such as biological oceanography, chemical oceanography, geological oceanography and physical oceanography.

4. Oceanography is significant to geography because the fields have overlapped in terms of navigation, mapping and the physical and biological study of Earth's environment—in this case the oceans.

5. Ocean waves can be caused by the wind, by earthquakes, and by other underwater phenomena.

Ⅲ. 1. 海洋学家 2. 大堡礁 3. 洋流 4. 海底扩张 5. 地质海洋学 6. meteorology 7. biological oceanography 8. kelp forest 9. coral reef 10. tsunami

Ⅳ. 海洋/大气相互作用是海洋学中另一个研究领域,主要研究气候变化和全球变暖之间的联系,以及由此引起的对生物圈的关注。大气和海洋主要是由于蒸发和降水联系在一起的。此外,像风这样的天气模式驱动着洋流,使不同的物种和污染物四处移动。

Text B Ocean Waves

1. Waves are the forward movement of the ocean's water due to the oscillation of water particles by the frictional drag of wind over the water's surface.

2. The wavelength, or horizontal size of the wave, is determined by the horizontal distance between two crests or two troughs. The vertical size of the wave is determined by the vertical distance between the two.

3. There are four different kinds of waves: wake, large groups of wave trains with enormous energy, tsunamis and swells.

4. When the waves become too high relative to the water's depth, the wave's stability is undermined and the entire wave topples onto the beach forming a breaker. Breakers come in different types—all of which are determined by the slope of the shoreline.

5. Headlands composed of rocks resistant to erosion jut into the ocean and force waves to bend around them. When this happens, the wave's energy is spread out over multiple areas and different sections of the coastline receive different amounts of energy and are thus shaped differently by waves.

Unit 18 Environmental Technology

Text A How to Stop Humans from Filling the World with Trash

Ⅰ. 1. T 2. F 3. T 4. F 5. T 6. T

Ⅱ. 1. Two ways: Charging money and burning garbage.

2. Pneumatic tubes will whisk trash to a sorting area.

3. Dioxin emissions from incineration plants caused cancer and birth effects.

4. Plasma gasification, an experimental technique, could eventually replace incineration as an even cleaner and more sufficient way to get rid of trash.

5. If manufacturers are required to fund and manage the recycling of their goods by the extended-producer-responsibility laws, it will give companies an incentive to make their products last longer.

Ⅲ. 1. 射频识别 2. 堆肥 3. 面部识别软件 4. 回收箱 5. 焚化厂 6. labor intensive 7. skyscraper 8. waste management 9. pyrolysis 10. non-recyclable materials

Ⅳ. 斯塔利说,如果火箭技术得到改善,终有一天,我们可能会把垃圾扔进太空,并利用太阳的热量将其燃烧。但是考虑到地球的资源有限,使用一次后燃烧它们可能并不是解决办法。一些环保主义者首先要阻止公司制造不可回收的材料,还有一些人提出了替代方案。一个欧洲研究小组设计了一种由可回收材料制成的笔记本电脑,其组件可以重复使用。

Text B Surviving on Earth

1. They need to cool the planet down and remove carbon dioxide from the atmosphere. These projects fall into two categories: solar management and carbon-dioxide removal.

2. The stratosphere is a layer of atmosphere that sits between 6 and 29 miles above the planet, where they reflected enough sunlight to lower global temperatures by 2. 2 degrees Fahrenheit on average.

3. Because the stratosphere is too high for rain to wash them out. These particles might remain floating in the stratosphere for up to two years, reflecting the light and preventing the sun from heating up the lower levels of the atmosphere where we live.

4. ① Using one of the earth's most adaptable organisms: algae called diatoms.

② Enlisting the acid of rocks.

③ Enhanced weathering.

5. Intense weathering from wind and rain wore the Appalachian Mountains down to a flat plain; runoff from the shrinking mountain took tons of carbon out of the air, raising oxygen levels and sending the planet from greenhouse to deadly icehouse.